TECHNOLOGY PRICING

Technology Pricing

From Principles to Strategy

Francis Bidault
Faculty Member, IMI–Geneva
Professor, Lyon Graduate School of Business

Translated by
Brian Page and Peter Sherwood

St. Martin's Press   New York

© Francis Bidault, 1989

All rights reserved. For information, write:
Scholarly and Reference Division
St. Martin's Press, Inc., 175 Fifth Avenue, New York, N.Y. 10010

First published in the United States of America in 1989

Printed in China

ISBN 0–312–02391–X

Library of Congress Cataloging-in-Publication Data
Bidault, Francis, 1949–
Technology pricing.
Bibliography: p.
Includes index.
1. Technology transfer—Economic aspects.
2. Pricing.   I. Title.
HC79.T4B53   1989   658.8′16   88–18156
ISBN 0–312–02391–X

To Caroline, Frédérique and Sébastien, who, without knowing it, have encouraged me to pursue this research to its conclusion

Contents

List of Figures

List of Tables

Acknowledgements

This book would certainly never have been completed without the help, encouragement and advice of a great many people to whom I wish to express my most sincere gratitude.

Although I must accept full responsibility for all imperfections and errors contained in this work, I must acknowledge the benefit I derived from discussions with Farok Contractor, Stephen Magee, Tagi Sagafi-Nejad and Howard Perlmutter from the very beginning of my reflections on this subject.

My students at the Lyon Graduate School of Business (ESC, CESMA, DEA in Management Sciences) and at the Institute of Political Science in Lyon have also enabled me, by their questions, to progress in the formulation of some of the points dealt with in this work. I would like to thank them, especially André Zervudachi and Eva Mullor with whom discussions developed into real team work.

The management consultants I met during my research have greatly helped me in defining the problem. I particularly wish to thank Roland Fabre, Jacques Robichon, Jean-François Tronchon and Maurice Vaucher.

I must also mention all those who helped me with the most strenuous phase of the project: the survey carried out in companies. This could not have been done without the financial support of the Fondation Nationale pour l'Enseignement de la Gestion des Entreprises (FNEGE), and the Association de Recherche en Propriété Industrielle et Transfert de Technologie (AREPIT). I express my gratitude to Michel Bernard of FNEGE and Gilles Bertin of AREPIT.

One difficult aspect of surveys is to identify the companies to be interviewed. An introduction by a person of note makes this much easier. This part was played by Bernard Prugnat, President of LES France, Vitor Corado Simoens for Portugal, and Mohamed Bendjillali of ODI for Morocco. But their help was not limited to this, and they also made some valuable suggestions. For all they have done I thank them most sincerely.

In France, the survey was carried out by Philippe Hug de Larauze, Yves Theron and Eva Mullor. Their personal commitment explains in great part the quantity and quality of the data obtained. I wish to thank them.

I also have in mind the many company managers and executives who were prepared to give of their time to answer our interviewers' questions. Without them, of course, this research would have been largely incomplete. I do hope that these results will help them in their work as negotiators.

The quality of the word processing for the original French text by Valérie Buclon and Nadine Pic also deserves a mention. They managed to lay out my manuscript, in spite of countless modifications and suggestions. I wish to express my gratitude to them. On the subject of these modifications, I must mention the contributions of Françoise Doucet and Paul Laurent who went through an initial version of this work most thoroughly. Their insight enabled me to avoid making mistakes of form, and their suggestions helped to improve the development of some of the ideas. Professor Claude Mouchot's apposite remarks were very valuable for maintaining a critical and objective attitude as regards the statistical results.

This sort of research often interferes with one's professional life. I am conscious of the fact that there were times when I was not sufficiently available to my colleagues Smaïl Aitelhadj, Christine di Domenico, Danielle Dubujet, Christiane Roche and Henri Tariel. But I was most receptive to their encouragement and advice, and I wish to thank them.

I would also like to thank my translators, Brian Page and Peter Sherwood, and the team who worked with them to produce this English-language version of my original French text: David Bailey, Alison Coulavin, Alain Guihard and Margaret Page.

The Lyon Graduate Business School has been a unique work environment because of its human and documentary resources. This work would have been difficult, or even impossible, anywhere else. I would like to express my most sincere gratitude to the School management for allowing me to carry out this project.

Lastly, I must mention the part played by Professors Roland Perez and Charles A. Michalet throughout this research. They were able to observe the slow, and sometimes chaotic, development which brought about this result. I was very honoured that they were willing to direct my research.

FRANCIS BIDAULT

'I do not believe that scientific progress is always best advanced by keeping an altogether open mind. It is often necessary to forget one's doubts and to follow the consequences of one's assumptions wherever they may lead – the great thing is not to be free of theoretical prejudices, but to have the right theoretical prejudices. And always the test of any theoretical preconception is in where it leads.'

Steven Weinberg, *The First Three Minutes of the Universe: A Modern View of the Origin of the Universe* (New York, Basic Books, 1977, p. 115)

Introduction

'There exists no standard method for determining a fair price for a technology', according to a report by UNIDO,[1] while the OECD experts state 'There is no insufficient or excessive price, there is a price, accepted by the two parties to the transfer'.[2]

These two assertions are a good expression of the opinion most commonly held by agents and observers of 'technology transfer'. Notwithstanding the various formulae used by specialists to determine the price of a technology, the general feeling is that this price depends in fact on the 'bargaining power' of the partners.

Indeed, while no 'standard method' exists, there are a fair number of 'recipes', most of them developed by international experts or consultants. They show which variables should be taken into account in evaluating prices, and how the calculation should be made. But quite considerable differences can be found between them, which cannot be explained. Hence the impression that these recipes are *arbitrary*, that 'fair price' is a myth, and that it is only 'bargaining power' that has any effect in the negotiation.

A certain suspicion as regards the methods seems to us to be justified, but the importance given to the bargaining power is merely a measure of our ignorance. Power is not acquired in its own right, it relies on supports, which have to be identified; to assert the importance of bargaining power without trying to find out the limits within which it operates, and which factors contribute to its development, is really admitting an inability to seize the problem.

A Question Neglected by Specialists

Indeed, this question has not attracted much attention on the part of technology transfer specialists who, apart from a few brilliant exceptions, have preferred to address themselves to other aspects of this phenomenon.

Yet there is no lack of 'resources', as over 2000 books dealing with technology transfers have been identified.[3] However, most publications concentrate on the *consequences* of the phenomenon, rather than on the *reasons for the agents' behaviour*, contrary to what happened during the multinationalisation of companies.[4]

Two research scholars specialised in this area, Farok Contractor and Tagi Sagafi-Nejad, in a remarkable synthesis, have listed the publications dealing with the subject, classifying them under a dozen headings. First of all they define six theories for discussion:[5]

- The *role of technology* in economic growth and development.
- The *content of technology transfer*, what influences it, and especially its effects on the relationship between the partners.
- The *effect of the international system of industrial property* on the diffusion of technology, particularly towards the developing countries.
- The *terms of contract* (from licensing agreements to 'joint ventures' including turn-key contracts), their justification, and their impact on the drawing of the rent.
- *Costs and payments*, essentially from the angle of the possibilities of government intervention in the recipient countries. Only a few publications dealing specifically with price determination are mentioned.
- The question of the *choice of technologies*, and their compatibility with the local environment; this question is raised at government level (which technology for what development) and at company level (adapting production to local conditions).

These two authors then give the political answers provided to these questions by the agents of technology transfer: companies, government and international organisations.

It appears from this albeit 'incomplete' survey[6] that one dimension of the phenomenon has been of particular interest to observers: the role of developing countries, and possible regulatory policies. The attention paid to these aspects is out of all proportion to the importance of developing countries in the international flow of technologies. On the other hand, the transfer of technologies towards industrial systems which are basically different from those where they were developed obviously raises complex problems and deserves special attention.

In fact, it can be argued that the success of the Japanese model of technological development has already added a new contribution to their discussion. Japan, whose technological success can no longer be questioned, has indeed applied a policy of technology import which many specialists consider to be a model, even though its application in other countries may face unavoidable obstacles.[7]

A Disturbing Question for the Decision Makers

Without denying the importance of this discussion, we find it surprising that the question of pricing has attracted so little attention over the years, particularly as all the agents of 'technology transfer' show great concern for the subject, and the conversations we have had with some of them have clearly shown that there is a need for clarification. This need is expressed today by statements, which are quite general in nature, which we would like to summarise in order to show how considerable the concern really is.

On one side we have the less developed countries – or rather their spokesmen – who regularly criticise the conditions under which they purchase technologies. They consider that their lack of bargaining power results in their paying more than the advanced industrial countries.[8] This point of view is sometimes supported by researchers who explain this situation by the shortage of financial resources, and of technical and marketing skills in the developing countries.[9]

But their claim concerning the cost of technology goes much further. Some people assert that technologies are part of the common heritage of mankind, in the same way as culture, and that consequently no one has any right to their ownership. Paul Marc Henry declares[10] 'The accumulated wealth of scientific knowledge and technological capability belongs to humanity as a whole and cannot be monopolised by the richest and most developed fraction'.

Such points of view can still be heard in most international forums where technology transfer is discussed. They challenge one of the basic principles of the developed economies which supply the greater part of technologies: the notion of intellectual property.

Representatives of the advanced industrial countries on their side strongly challenge this approach by claiming that the level of prices is justified, and that any attempt to interfere would only discourage these countries from supplying technologies. This, in turn, would result in a decrease in the flow of technologies towards developing countries. This response has been confirmed by several researchers. For instance, Mason claims that multinational firms apply lower royalty rates to developing countries than they do to advanced industrial nations, and that consequently it cannot be stated that these prices are too high.[11]

Following the same line, Teece asserts that present practice concerning the transfer of technologies, including restrictive clauses, are necessary to provide adequate incentives to exporters.[12]

But disagreements on the price of technologies are found not only in discussions between states. Contrary to Mason's assertions, and in spite of their being based on a survey carried out in Mexico and the Philippines,[13] it often happens that the licensee himself considers that the price he is paying is too high.

A study carried out by Baasche and Duerr among suppliers and recipients of technology reveals that the price is the first source of conflict between them.[14] The former, whom we will refer to as licensors, consider their remuneration to be fair (perhaps even slightly low), considering the cost of developing the technology, and the cost of the associated services. They point out that in addition, they enable their clients to make considerable savings in time and money by not having to develop these technologies themselves. The licensees, on the other hand, accuse their partners of taking advantage of their dominant position, forcing them to make tied purchase which increases the cost of acquiring the technologies. They regret not being better informed about the various sources of supply. It seems that there is insufficient competition in the 'technology market'.

However, it has to be recognised that the problem does not present itself in exactly the same terms for the four agents in technology transfer: the recipient state, the supplier state, the supplier firms and the recipient firms. For the states, it is essentially a question of controlling the balance of basic macroeconomic variables because these operations have a considerable effect on the rate of growth, employment, the balance of payments, and ultimately the nation's independence.

On the other hand, companies which carry out these transactions are concerned essentially with microeconomic considerations, which means that their interests are not always compatible with those of their respective states. Perlmutter and Sagafi-Nejad[15] suggest a synthetic vision of this 'quadrilogue' on price where disagreements are seen to arise between companies and their respective states. But probably more interesting is the conclusion drawn from the results of the survey among representatives of the four groups – i.e., that there is fairly broad agreement on the wish to define guidelines in order to make the evaluation of price more objective. The need to clarify the pricing of technologies seems to be generally recognised.

Such a clarification has to be made at company level, inasmuch as it is companies who hold the information on the variables which affect the determination of prices. This methodological requirement also partly explains the absence of research in this area, because access to information is made difficult for two essential reasons. First, most of the required data exists only in an informal manner inside the companies, as few of them keep separate accounts for their licensing agreements.[16] So the researcher often has to carry out in-depth surveys among company executives. Secondly, even if the data were available, the problem of 'confidentiality' would arise. This difficulty can quite often be dissuasive, and can be overcome only by building up a relationship based on trust. This is what we have endeavoured to do.

A New Approach to the Question of Technology Pricing

The ambition of this book is to explain how the price of technologies is determined, and to suggest improvements on existing practices.

In other words, we wish to make proposals to the agents of technology transfer, thereby adopting an action-oriented approach, which goes beyond the phase of strict observation. However, neglecting concrete behaviour on the grounds that it is imperfect could lead us to formulate unrealistic recommendations. Hence our study will take the form of an empirical analysis of a sample of licensing contracts in which we will attempt to identify the essential price determinants.

In carrying out this task, we must admit that, in spite of the limited number of works devoted to this question, we are not starting from scratch. Accounts and results of surveys are already available, and we will have to judge their relevance, and also their limits.

We will draw attention in particular to one aspect which has been underestimated and even neglected in earlier publications: the *strategy* followed by the two partners in the licensing relationship. All licensors do not have the same motives for transferring their technologies, and similarly, there are differences among licensees. We believe that the strategic dimension is an interesting starting point for a better understanding of a firm's behaviour, and for correcting the approximate methods which they may use for fixing their prices. This leads us basically to discard the 'economist' viewpoint which has prevailed up to now, which led people to believe that there was such a thing as a single and fair price.

But our ultimate goal is to suggest *models* for the determination of price, thus putting into concrete form our action-oriented approach. These are 'models' because our intention is to tell the agents 'what they should do' but, contrary to the methods in use today, we envisage several cases depending on the strategy being followed. So the construction of these models will form the invisible texture of this work, and their presentation will appear in many ways as a synthesis of the contributions which preceded it.

We also want this book to be of some use to the agents of 'technology transfer'. First to the licensors, who should find in it recommendations intended to improve the coherence between their pricing policy and their strategy. Also to the licensees, to whom we recommend an analysis of their partners' motives, to see if these seem acceptable to them before defining their attitude in the negotiations. Our observations have been based solely on transfers of technology between firms having no financial relationship with each other, for reasons of methodology which we will explain later. However, we believe our results could also be of interest to firms transferring technologies to their subsidiaries abroad. We define rules of evaluation which should enable them to clarify their operations in the eyes of local governments, but also of the managers of subsidiaries who sometimes feel that they are paying unjustified royalties which have a negative effect on their results. Lastly, let us hope that governments of supplier countries, as well as those of recipient countries, will find in this work food for thought on the role they should play. We particularly have in mind the regulatory organisations which are responsible for the approvals without which the contracts cannot come into effect. They should benefit here from a series of methods which are less normative than those which are presently being used, because licensing relationships are still too often regarded as a homogeneous phenomenon.

But, before introducing the reader to the analysis of technology pricing, it is necessary to define carefully the subject of the discussion. Technology transfer is such a vast area, with so few landmarks, that the risk of confusion is considerable. Hence Chapter 1 will be devoted to giving the necessary definitions, drawing the limits of our problem, indicating the essential features of our approach and the methods used to validate it. Only at the end of the chapter will we present the main stages in the development of this work.

1 International Technology Transfer: Substance and Framework

The popularity of the expression 'technology transfer' is equalled only by its *ambiguity*. The expression comes up in the most diverse contexts. It sometimes refers to the transfer of know-how from one company to another, sometimes to the establishment by a multinational corporation of a manufacturing facility in a particular country, sometimes to the sale of capital goods with the provision of related services, sometimes to technical training programmes, sometimes to the link between the development and the manufacture of a product ... This list, although far from complete, gives an indication of the great variety of transactions covered by the expression.

It is therefore important to define the field to which our pricing analysis refers, because the problems encountered obviously vary considerably between one type of transaction and the other. We will explain in this chapter that we are concerned only with the transfer of technologies – i.e., a body of knowledge and proprietary rights – between independent firms engaged in a similar industrial and commercial activity. This field of activity appears to us as the most worthy of study, for it is probably in this area that the agents are most lacking in precise information. Moreover, a contribution in this field will no doubt open up opportunities for further study in others.

Our approach will centre around two dimensions: the *object* and the *framework* of the exchange. In the first place, we will define what we mean by 'technologies', and we will draw some conclusion concerning the notion of 'price'. Secondly, we will define the type of partners, and the terms of contract to which we will be referring.

Then, thirdly, we will explain our approach to pricing, and emphasise the necessity of linking it to corporate strategy. Finally, we will describe briefly the methods used for the research on which this book is partly based.

1.1 THE SUBSTANCE OF EXCHANGES: TECHNOLOGY AND KNOWLEDGE

Technology can be defined as a body of processes or methods which are used to produce goods. We are therefore talking about the practical knowledge which is necessary to manufacture a product.

But the word 'technology' refers to several notions which, from the point of view of their transfer, are very different one from the other: not all technologies can be transferred. We will specify which types we will select. Then we will examine to what extent one can talk of the market, the sale, the price of a technology.

The Different Types of Technology

What we mean by 'technology' can be defined by using four concepts: appropriation, the terms of appropriation, the channel, and the ultimate purpose of the process.

The Degree of Appropriation of a Technology

Technology can take two forms of appropriation: it is either public or private.

A technology is public when it is *freely accessible*, with no restrictions. This is usually the case when the knowledge is widely and openly available, so that private appropriation has become impossible. In this case, the recipient does not pay for the knowledge (which as a rule has no value), but only for the medium of transfer (as in the case of general technical training).

Conversely, a technology is said to be private when *access is restricted* by proprietary rights (patent, model), or by a specific agreement. This concerns the technical knowledge of which a company is recognised as the owner, and that which it has managed to keep secret.

An analysis of this distinction, due to Gonot,[1] reveals that the boundary between the two statuses of technology results from a difference of degree rather than of nature. But, more importantly, this degree of appropriation depends largely on the strategy followed by the transferor who can, to some extent, delay making his technology public.

The distinction between public and private technology may be

possible diffusion, this is seldom the case with disembodied, or human-embodied technology.

The Ultimate Purpose of the Process

Technical knowledge can be broken down acccording to whether it applies to the use of a product, or to its manufacture.

The correct use of a machine sometimes – and in fact increasingly often – requires that the buyer receives training in its use or operation. This in itself can be considered as a form of 'technology transfer'. The same can apply to consumer goods. One can think of home computers, with their ever-increasing capabilities, which create training requirements for the user.

But, taking into account what has been said earlier, we will exclude these transactions from technology transfers in the strict sense. In other words we will talk of technology transfers only if they are intended to be used by the recipient for the manufacture of products.

Finally, we will conclude from all these considerations that our subject will be limited to private disembodied technologies, which may or may not be the subject of patent rights, and which are used for the production of goods. According to this concept, technologies appear essentially as a body of rights and data. This observation leads us now to point out some of the implications concerning their price, in the light of the theory of information.

Implications for Price Analysis

The explanations which follow are not intended to cover the question of pricing, but to underline the impossibility of assimilating the price of a technology to the price of a product. We will see that while technology is indeed exchangeable, it does not have the characteristics which are essential to goods, and follows specific laws in the definition of its value.

Technology is Exchangeable Because it is Private

A basic characteristic of technology as information is that it is by nature a 'public good'. This concept defines a product which, while it is being used by one person, can be used simultaneously by other people. Information is inexhaustible, because it is not destroyed by use. It follows, as Arrow[3] explains, that the optimum allocation of resources implies that this 'product' is available free of charge. This theoretical requirement runs against the necessity for states to encourage companies to invest in research and development. It is also likely that if the information were available to all competitors, the *return* on investment for the innovator would be *inadequate*. Thus, in order to encourage technical development, states are forced to intervene, therefore recognising 'market failure'. They intervene in particular at two levels where innovations are made by companies. On the one hand, they grant a temporary monopoly by the use of patents. On the other, they allow impediments to the free movement of information by legalising industrial secrets.[4] We will come back to the logic of these government interventions in Chapter 3. We simply wish to point out here that the *private* nature of technology not only encourages innovation, but is the absolute condition for technology transfers to take place. 'If a thing cannot be a property, it obviously cannot be a commodity' writes Boulding.[5]

So technology is exchangeable only because it is private, and it is private, directly or indirectly, only because the state sees a failure in the operation of the market, with regard to the production of information. Other forms of organisation are certainly possible,[6] and we will present them later. But 'privatisation' remains the leading form in industrially developed countries which provide the greater share of technologies.

The concept of technology as a public good has often been misunderstood. Some commentators see it as an ideological standpoint aiming to justify the diffusion of technical information free of charge. This erroneous interpretation is found among writers who claim to defend the interests of the Third World, and takes the form, as we have seen, of a theory on technology as the common heritage of the human race. They do not realise that this claim completely neglects the basic interests of the suppliers of technology – at least within the present state of industrial property. But it seems equally illusory to question the nature of technology as public good, as other observers do, who want to justify the 'trading' of technology.[7] They criticise what they believe to be a biased attitude, while technology presents by its very nature the characteristics of a public good whose effects are hampered, deliberately, by action of the state, or of companies.

Technology is Not a Commodity

Having established that technology is the subject of exchanges is not sufficient to conclude that it follows the same laws of the market as any other product.[8] As information, technology offers a specific nature of supply and demand that *rules out any analogy* with products.

First of all technology is not quantifiable, cannot be counted, in a simple manner. Quantification, consisting in measuring the space occupied by a piece of information, is not adequate because it leaves out the *significance* of the information.[9] More precisely, there exists no measure of the quantity of information, independent of the use that will be made of it.[10] As a result, the value of information – the benefit which it affords – depends upon the buyer's perception, and therefore on the information he already holds. Consequently, one cannot draw a market demand curve as one can, at least theoretically, for products.

Furthermore, the cost of information – that is, what would form the base of the supply curve – follows specific rules.[11] Usually the reproduction cost of an item of information (communication to a user other than the owner) is far below the cost of initial production. Unlike the case with products, we would have a constantly decreasing marginal cost curve, even if the cost remains substantially positive. Moreover, the cost of using a piece of information, for the same user, is also decreasing. Technology has another important characteristic concerning its cost: it is often a *by-product*.[12] The production of information is not a linear process. Research and development programmes often lead to results with multiple applications, and it is practically impossible to determine the resources which have been used for one particular innovation. Nor does know-how result from a process which can really be isolated. It is the fruit of experience, of the accumulation of ideas of improvement, in a nutshell, of 'learning by doing': its production is a 'by-product' of a manufacturing activity. It is difficult to calculate the cost of a by-product; the same is true of technology, in most cases.

Besides, technology is *seldom intended to be sold*. There is a broad consensus on this point.[13] Most companies, apart from private research laboratories or engineering consultants which are outside the scope of our investigation, envisage the production of information only for their own needs. They then transfer these technologies for reasons of opportunity, or else because this can serve their strategic interests.

This means that drawing supply and demand curves would be completely artificial. There is an exchange, but there are no products. Consequently, we regard the expression 'technology market' as inadequate, inasmuch as it implies that there are buyers and sellers, supply and demand which should allow the definition of a point of equilibrium. This is not the case; technology transfer is a two-way relationship in an environment which may be competitive. Hence the implication that the value of technology differs fundamentally from traditional principles.

Technology Has a Use Value

Considering the characteristics of information and of technology in particular, the prevailing idea is to link the value of an item of information to the income it brings to the user. The value of a piece of information depends on the profit it generates, or the loss which it enables the user to avoid.[14]

One condition must be laid down: the value of information must exceed the cost of the channel or medium of information.[15] In other words, a technology can be sold only if the income it generates exceeds the cost of transfer. This concept is the principal yardstick for considerations on the pricing of technology. We will therefore refer to it when we go into the actual theory of technology pricing (Chapter 3), but we must first give some detailed indication on the notion of pricing, by presenting the different conditions under which the seller is remunerated. However we can point out straightaway that one of the main criticisms we have towards this approach is its lack of consideration for institutional phenomena and the behaviour of the agents. The result is that the effective price varies considerably from the technology's use value, insofar as the latter can be identified *ex ante*.

1.2   THE FRAMEWORK OF EXCHANGES: TRANSFER AND TRANSACTION

The term 'transfer', which could come from the theory of information, has a misleading connotation because in certain circumstances, it refers to a transmission without compensation – as, for instance, in the case of foreign exchange transfers made by foreign workers to their home countries.

Many authors[16] have criticised this expression, which conveys erroneous connotations, not only on the part of buyers who see it as a humanist dress-up for a mercantile transaction, but also on sellers who, for their part, want to assert the private nature of technology, particularly during discussions on the 'code of conduct for technology transfers'.

Other expressions have been suggested. For instance, Dunning would prefer to use the term 'transmission';[17] but to our mind, this does not convey the economic dimension of these operations. We would prefer to talk about the supply, even the exchange, of technology, avoiding where possible the word 'transfer'. But it must be recognised that its use is so widespread that we will more or less be forced to include it in our vocabulary, if only to refer back to previous publications.

So several types of technology transfers have been distinguished. We must point out which types we will be dealing with, after which we will indicate the terms of contract relating to this concept.

The Different Types of Transfer

The distinctions have been based on three dimensions: the business activity of the supplier firm, the nationality of the recipient, and his status.[18]

The Business Activity of the Supplier Firm

The supplier firm's activity is frequently the same as the recipient's; we refer in this case to horizontal transfer, because it takes place between two companies on the 'same level' of the vertical chain of activities. They are parallel, each in its own market.

There is another form of 'transfer' of technical information which takes place between Research and Development laboratories and production workshops. In other words, this is the transmission of information between the different stages which lead to the perfecting of an innovation. This process is described as 'vertical transfer'. The supplier firm is naturally positioned further from the market than the recipient. The opposition between *vertical* and *horizontal* transfer is now widely recognized.[19] It is clear that the nature of these operations is basically different.

We believe that this distinction can be taken further by identifying more precisely the suppliers of technology, which of course are not limited to these two extreme types. Hence, in an article dealing with the question of return competition, we introduced two other possible suppliers: manufacturers of capital goods, and engineering consultants (see Figure 1.2).[20]

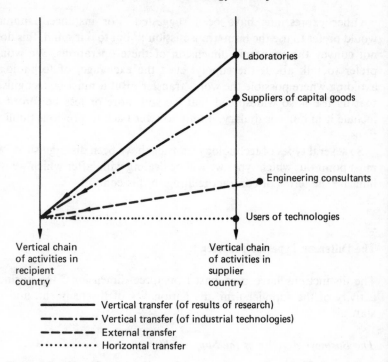

Figure 1.2 Classification of transfers according to the supplier's activity

Manufacturers of capital goods sometimes try to support their sales by providing technical information which goes beyond simple instructions for use. They may then contact their clients who have the effective know-how in an attempt to associate them in a 'transfer of technology' towards the future users of the machine. We refer to these somewhat unusual operations by keeping the expression 'vertical transfer', inasmuch as the movement operates in the same direction, but we must point out that they are industrial technologies, and not only the results of research.

Engineering consultants take an important part in the international flow of technologies, particularly as coordinators of turn-key units. These companies are often specialised by industry, and they are found particularly in the oil, chemical, and petrochemical industries. They come to provide technologies of which they are already masters, or which they acquire from industrial companies which apply them.[21]

These two types of suppliers, as well as laboratories, have motivations for the transfer of technologies which are very different from those of the users of these technologies. This is largely due to the risks of return competition, which effectively is run only by users. It can thus be expected that the pricing problem will present itself differently according to the type of supplier.

The Nationality of the Recipient

One can consider that, on the face of it, technology transfer retains the same nature, whatever the nationality of the recipient. Indeed, from a theoretical point of view, the important thing is to specify the geographical area in which the licensee will be able to make use of the technology he has acquired. The fact that this area does or does not coincide with national boundaries is of no consequence.

Teece has pointed out other differences of form which can be important.[22] International operations often require the *adaptation* of products or processes to the recipient market, and also additional costs arising from distance, differences in language, culture and environment.

It should be said, however, that the greater part of horizontal transfer agreements takes place on an international level, where the risk of competition with the licensee can be controlled more easily, or with less difficulty. The risk of competition on the licensee's part also depends on whether or not he is independent of the licensor. This brings us to examine this question, which has multiple implications.

The Status of the Recipient

The licensee is not always an independent company. In fact, it is often a subsidiary of the licensor, and the transfer of technology takes place between two legal entities which are under the same financial control.[23] These operations are called *internal transfers*, as opposed to those which take place between independent partners, which are known as *external transfers*.

It is useful to point out that internal transfers represent a significant part of the international flow of technologies. Although comprehensive statistics are not available, one can imagine the large volume of these transactions from information coming from the USA, the UK, and Western Germany.

From the statistics put together by Madeuf, it appears that in 1980 over two-thirds of technology payments were made by subsidiaries: 67.5 per cent of payments made by the USA, 84.8 per cent by the UK and 61.7 per cent by Western Germany.[24] Hence it is likely that the 'closed market' for technologies (inside multinational corporations) represents over 50 per cent (maybe over 60 per cent) of the world total. Furthermore, the trend seems to be towards an increase in this share.

In spite of the considerable volume of technology transfers within multinational corporations, we will not consider these transactions. They are not, in fact, 'actual transfers of technology',[25] inasmuch as the technologies continue to be appropriated through subsidiaries. One can therefore make the assumption that the conditions of transfer are largely different from those of the 'market' – i.e., those which prevail in the relationship between a licensor and a licensee which are financially independent one from the other.

The analogy with the question of 'transfer pricing' of products between a parent company and its subsidiaries[26] implies that the prices have been fixed according to specific fiscal or strategic criteria. The result is that there is so much bias in the pricing policies that it is impossible to use them as the basis for a factual study.[27]

In conclusion, we will speak only of technology pricing on horizontal, international, and external transfers. It is these operations which are the form of transmission closest to marketing. It is only for this type of transfer that the concept of technology pricing can make sense, owing to the many *interferences* which affect the others.

We therefore consider as technology transfers, within the scope of our investigation, essentially the transmission of information against payment, with the corresponding operating rights, between two companies with no relationship of financial dependence between them, of which one was or remains a user of the relevant technologies, while the other becomes a user, or remains one with added efficiency.

This specific form of relations between firms is a means of development which competes with other alternatives. We will have to *compare* their respective advantages and disadvantages in due course, particularly with reference to multinationalisation.

These preliminary matters having been dealt with, we now have to lay down in more practical terms, precise details concerning the terms of contract which we have selected.

Terms of Contract

In actual fact, technology transfers take the form of intricate operations which often combine several types of supplies. We can, like
Teece[28] distinguish the input of:

- *Intangible goods*: technical information, patent rights, etc.
- *Material goods*: capital goods, special tools, raw materials.
- *Services*: technical assistance, sales assistance, maintenance, engin-
  eering studies, training, etc.

These three types of input are not situated on the same level. Some are
essential for defining an operation as technology transfer – e.g., the
communication of information – while others – like the supply of raw
materials – are secondary. The latter are nonetheless mentioned,
because they frequently appear in transfers of technology. Moreover,
some inputs actually serve as the channel or medium for the others; the
delivery of special tooling is often a necessary condition for the effective
communication of know-how. This is why it is not possible to limit the
transfer of technologies solely to the input of intangible goods. They
form a combination of factors which vary in number from one case to
the other.

The most advanced form of transfers of technology, from this point
of view, is what is referred to as the sale of industrial complexes – i.e.,
turn-key, or product-ready contracts. These agreements usually involve
the three types of contribution simultaneously. For this reason, their
application raises complex problems, particularly in the legal field.[29]
But their importance should not be overestimated, even if their
complexity has attracted a great deal of attention. They certainly do
not hold a dominant position in the international flow of technologies,
because we know that developing countries – which are the main
buyers of these industrial complexes – account for less than 20 per cent
of the market.[30] The greater part of technological exchanges take place
between industrially advanced countries in operations which are
usually far less complex, most of which nonetheless combine at least
two of the three types of input, and therefore involve a number of
different benefits.

The Practical Identification of Technology Transfers

In order to determine which agreements come within the scope of our
investigation, we have made up a selective list of benefits which, in our

opinion, include the supply of technologies. To do this, we have referred to previous works on the subject.[31] The definition of benefits concerned with technology transfers varies considerably, because the borderline between what is essential, and what is complementary does not always follow the same course. On this point, we have to recognise that there is always a purely discretionary element.

We have therefore made a point of drawing a fairly long list of all possible benefits, using an easily understandable segmentation. On this list, we have identified a limited number of benefits which are in themselves inputs of technology, or which are indispensable as channels for this input. We consider as an input of technologies in itself, the transfer of rights and the communication of documents. Of course, transferring a right (patent or trade mark) may seem a purely formal operation, but it is in fact a necessary condition for the licensee to be able to use the information he has received: without the transfer of rights, there can be no real transfer of technology. The communication of technical literature refers to this information, which takes the form of drawings, plans, process manuals, manufacturing instructions (specifications, production schedules, engineering, tests).[32] But the communication of know-how requires, in most cases, the resort to services which form, as we said earlier, the channel for the transmission of information. This refers to the design and study of installations (or integrated consultancy), technical assistance, sales assistance, assistance for the maintenance and even for the sale of special tooling (moulds, dies) which include part of the knowledge being transferred. The sale of 'turn-key' factories falls within this type of service because, like them, it forms a channel for technology transfers, obviously in a more integrated form.

As for the other services – i.e., those on the list in Table 1.1 which do not have an asterisk – they are for the most part associated services which, taken individually, cannot be considered as technology transfers. This is the case with the sale of capital goods, the sale of intermediary products, the buy-back of products, and assistance in recruiting staff. We have also considered that training, and a minority holding in a joint venture, were not sufficient to qualify as technology transfer. As far as training is concerned, although it is found in nearly all operations, it is not in itself a transfer of the appropriate technologies, otherwise all courses within companies would qualify as transfers of technology. The reasoning is similar concerning a share in the licensee's capital. It is quite often a condition necessary to the

Table 1.1   The terms of contract for technology transfers

Service	Licensor sample (33 contracts) (%)	Licensee sample (29 contracts) (%)
1* Technical assistance	84.8	86.2
2* Sale of plans, technical manuals, tooling	81.8	93.1
3* Design and/or study of installations	45.5	41.4
4 Training of partner's staff	78.8	69.0
5 Turn-key sales of equipment (workshops, assembly lines)	36.4	17.2
6* Transfer of patents or licences	72.7	72.4
7* Transfer of trade marks (franchising)	24.2	48.3
8 Induced sales of intermediary products	51.5	48.3
9* Assistance in the maintenance of installations	24.2	6.9
10* Sale of turn-key factories	9.1	0
11 Minority holding in a joint venture	18.2	13.8
12 Buy-back agreement for partner's product	6.1	20.7
13* Sales assistance	21.2	20.7
14 Hiring of staff for partner	6.1	3.4

Note: Only the services marked with an asterisk are considered to be technology inputs, the others being complementary inputs.

co-operation, but it is never sufficient because a financial contribution does not necessarily entail an input of technologies.

Table 1.1 shows the complete list of services. It also indicates their frequency in the two surveys we refer to throughout this book.

The Part Played by Rights in Transfers of Technology

These data confirm a fundamental result, which is that the transmission of rights (patents, licences, franchises) is not always included in contracts for the transfer of technology.[33] It appears in about two-thirds of the contracts, whereas the communication of know-how through technical assistance, and the sale of literature, is far more widespread. This leads to two conclusions: firstly, rights are *seldom transferred separately* from technology, because the licensee would have to hold the technical data already; secondly, the technologies being transferred are not necessarily protected by industrial property rights. This underlines the part played by the transmission of information with regard to the simple transfer of rights: the latter cannot take place without the former, but the opposite is not true.

In any case, one finds a considerable variety of services being provided, and it seems practically impossible to describe these operations by one generic term. However, for practical purposes, we will sometimes use the term 'licensing' which describes not only temporary transfers arising out of a patent, but more generally the whole of the transaction between the supplier and the recipient of technologies.

The International Flow of Technology

It is regrettable that, in spite of considerable research,[34] the *volume* of the international flow of technology remains uncertain, even though a few authors have succeeded in carrying out a limited estimate of the phenomenon.

Estimates have been made, like those quoted by Janisewsky and Besso, which state that worldwide technology payments have risen from US$ 2.5 billion in 1965 to US$ 11 billion in 1975.[35] Perlmutter and Sagafi-Nejad estimate that these payments could exceed US$ 15 billion in 1980.[36] But these evaluations are not consonant with those published elsewhere, for example by OECD. Differences can possibly be explained by the inherent difficulty of statistical work in this field.[37] The fact remains that any hasty conclusion should be avoided. We can do no more than underline a number of points on which most specialists agree:

- Technology exchanges take place mainly between *industrially developed* countries.
- The share of internal exchanges within multinational corporations is *dominant*, and is apparently continuing to increase.
- Developing countries offer *more and more opportunities*.
- These countries are even beginning to *export technologies* (India, Brazil, Korea, among others).

The distribution of technology payments among industrially advanced countries (Table 1.2) gives an overview – certainly incomplete, but

Table 1.2 Balance of technology payments (million 1975 dollars)

	Year 1972						Year 1982					
	Revenues	(%)	Payments	(%)	Balance	Cover rate	Revenues	(%)	Payments	(%)	Balance	Cover rate
	(US$bn)		(US$bn)		(US$bn)	(%)	(US$bn)		(US$bn)		(US$bn)	(%)
USA	3214.8	66.7	368.3	10.0	2846.4	873	4130.9	63.3	202.4	5.4	3928.5	2040.9
Japan	226.5	4.7	934.0	25.5	−707.5	24.2	526.5	8.1	804.7	21.6	−278.2	65.4
Germany	270.4	5.6	631.6	17.2	−361.1	42.8	342.1	5.2	679.5	18.3	−337.4	50.3
France	302.8	6.3	460.8	12.6	−158.0	65.7	552.4	8.5	644.2	17.3	−91.9	85.7
United Kingdom	563.9	11.7	511.0	13.9	52.9	110.3	610.9	9.4	463.6	12.5	147.2	131.8
Italy	81.3	1.7	472.8	12.9	−391.5	17.2	133.9	2.0	498.6	13.4	−364.8	26.9
Holland	151.3	3.1	223.1	6.1	−71.8	67.8	210.0	3.2	352.7	9.5	−142.8	59.5
Austria	12.0	0.02	63.8	1.7	−51.8	18.8	20.9	0.3	73.0	2.0	−52.1	28.6
Total	4823.0	100	3665.4	100	1157.6	131.6	6527.6	100	3718.7	100	2808.5	175.5

Source: Madeuf (1985) p. 9.

nonetheless enlightening – of this phenomenon. One notices the progression in the flows, and the marked disparity between different countries.

As for stating how much 'external' technology transfers represent on an international level, this can only be an approximation. The best we can do, using the information published by a few countries, is to consider that these flows make up 40 to 50 per cent of the world total, which probably means US $4, 5 or 6 billion, according to the statistics recently published by Madeuf.[38]

So we are fairly sure that the phenomenon we are dealing with is not marginal, at least from a quantitative point of view. But in fact the actual importance of the phenomenon does not only lie here, but in the tremendous stake which a technology deal represents for a company. As a licensor, it may draw considerable income from it or, on the contrary, it may waste its technological advance. As a licensee, it can achieve a decisive technological progress, or fall under the dependence of a more powerful partner. These challenges make us aware of the strategic dimension of technology transfers, this dimension having often been underestimated in the discussion on pricing.

Now that we have clarified the essential terms of the problem, we can present the approach we have followed towards its resolution. The end of this introductory chapter will therefore be devoted to explaining how we treat the subject, and the method we have used to test the validity of the hypothesis. The chapter will conclude by a presentation of the general plan of this work.

1.3 TECHNOLOGY TRANSFER, CORPORATE STRATEGY AND TECHNOLOGY PRICING

The analysis of factors which influence the pricing of technology transfers has already been the subject of several publications. But there are not that many, as we have pointed out, considering how critical this question is.

The authors who have written on the subject have not all followed the same investigative approach. Some have made purely *theoretical* investigations which quite often place the pricing of technologies in the neoclassical microeconomic framework, thus assimilating technologies to products. In other cases, the approach has been essentially *empirical*, with the inventory and evaluation of the hypothetical factors determining the price. Lastly, a third group of works merely suggests, in a *normative* form, without any real attempt at validation, methods for calculating the price of a technology. Unfortunately, it is most unusual to find these three facets of the analysis (theoretical, empirical, normative) being developed and clearly set forth in a single definition of the problem.

The Strategic Dimension: Frequently Ignored

We will analyse these contributions in the first part of the book, noting their interesting features, and their limitations. Their contribution is obviously valuable in understanding technology pricing. But a critical analysis will show that their essential limitation comes from their being unable to take into account the motivations underlying the transfer of technical knowledge. In short – and this point will be developed extensively – most analyses tend to accept that the transfer of a technology is entirely motivated by the profit that can be derived from it. The seller of a technology is considered to be like the seller of *any kind of product*. Consequently, these analyses refer implicitly to the traditional rules of the market for goods and services.

Now, as we know, market theory is based entirely on the postulate that the objective of any business is to maximise profit – i.e., to create a surplus between the income derived from the sale of a product and its cost. This assumption is present in most schools of thought. But, even if this is a debatable point, we will not go into a discussion of the function or objective of a business undertaking.[39] And if we accept this assumption for the time being, it does not follow that it necessarily applies to the transfer of technologies.

For this to be true, the technology would need to have, in relation to the transfer, a status equivalent to that of a product. This would mean, in practical terms, that the sale of technological knowledge is the object of the company's activity. But we have seen that technology most often has the status of a by-product with, in addition, the specific characteristic of being unmeasurable. And we know the complications created in economic theories by the presence of by-products: many results obtained in single production have to be abandoned.[40] It thus seems difficult to treat licensing agreements simply as commercial transactions, except if we consider only those special cases where the production and sale of technologies is the company's professional activity. In other words, we refuse to regard the transfer of a technology as an act having as its sole justification the opportunity of making a profit.

Licensing: A Strategic Decision

We believe, on the contrary, that this decision often takes on a *strategic character* for the company, inasmuch as it can have a durable effect on its future. Indeed, while most authors consider implicitly that profit is the only motivation for technology transfer, our own observations contradict this view. We will see that technology transfer often serves the global strategy of a company, especially when it is used to strengthen its basic activity. Let us therefore begin by looking at the strategy of licensing arrangements. This will enable us to distinguish *three basic strategies*, which will be the framework for the study of pricing.

The pricing policy defined by the negotiator does indeed depend very much on his objective. We would not expect to find the same price where the aim is to maximise profit, and where it is to create a special commerical relationship with the licensee. In the latter case, we can expect the seller to make a special effort in the hope of promoting the achievement of his objective.

Several authors have already pointed out that pricing is strongly influenced by the licensing strategy adopted by the supplier firm. It will suffice here to quote from the short book by R. M. Bizec: 'The strategic aspect is more important for pricing than the technology itself'.[41] Similar views are held by Enid B. Lovell, Alain Weil, and the secretariat of UNIDO.[42]

The Link Between Licensing Strategy and Pricing Policy

But until recently, no book has addressed itself entirely to the influence of licensing strategy on technology pricing. In fact, few works really deal with this question at all. We do, on frequent occasions, come across incidental remarks on the theoretical price or the cost of technologies. One seldom finds constructive proposals on this problem.

Our project therefore consists in finding the *link* between licensing strategy and the price of technologies. This entails, to be more precise, an analysis of the *effects of the licensor's strategic options on his pricing policy*. In other words, we will try to answer the question: 'To what extent does a technology's price level depend on the strategy being followed by the seller?' Our intention is not to neglect previous contributions, but rather to place them in a wider perspective, taking into account all the factors determining a price which have been identified, and to show how they operate when the various strategic objectives of licensing agreements are introduced.

Finally, we will endeavour to define concrete recommendations for negotiators, from the vast quantity of data available, and from our own results. More specifically, we will present in the last chapter a series of *models* making it possible to define a *licensing strategy*, and to determine the corresponding *pricing policy*. We will point out, as we proceed, what each step has contributed to the final models.

1.4 METHODOLOGY AND OVERVIEW OF THE BOOK

In order to make sure that a relationship does exist between the strategy followed and the pricing policy of the licensor, we have assembled some qualitative and quantitative data relating to a number of licensing agreements. The results of this survey appear throughout this book, to support our argument. It is therefore useful that we explain how they were obtained.

We must emphasise at the outset that we did not restrict the survey to licensors only. Firstly, because our investigation was also aiming to redefine the part played by the other factors determining the price which were already identified. So restricting the survey to a sample of licensors would probably have introduced a bias in the analysis, insofar as the seller does not hold all the information required. Furthermore, it seemed to us that if the licensor's strategy did have an effect on pricing, then it should also be possible to analyse it through information held by the licensee. In a way, the survey among a sample of licensees was to enable us to reach similar conclusions in order to strengthen our empirical validation.

It is important to stress that we deliberately avoided questioning pairs of licensors and licensees on contracts between them. The reason is that such a procedure would have brought more disadvantages than advantages, however considerable the latter may have been. Indeed we had to collect information on objectives, prices, costs and profits, which are by nature extremely confidential. And we considered that if our interviewees had known that we were also going to meet their partners, they would have feared that we would disclose some of this information – e.g., the profits made on sales derived from the agreement. So this quest for symmetry would have been carried out at the expense of the quality (and even the quantity) of data obtained. We therefore preferred to construct *two unrelated samples* (one for licensors, one for licensees) in order to reassure the firms we questioned, whose trust we nonetheless needed to obtain.

We undertook with them not to disclose any individual data; the presentation of the results does take this restraint into account. This pledge enabled us to obtain a remarkably low rate of non-response: out of 70 contracts which were the subject of an interview, we were able to process 62 questionnaires, filled in by 51 companies.[43]

The questionnaires divide into two samples:

- The licensor sample, made up of 33 contracts signed by French companies with partners in various countries.
- The licensee sample, with 29 contracts signed by French, Portuguese and Moroccan companies (respectively 7, 8 and 14 contracts).

In contrast with earlier studies, we endeavoured to contact not only big companies, including multinationals, but also small-to-medium ones (i.e., those employing less than 500 people, which is the usual definition internationally). The latter group accounted for over a third of the companies contacted in the licensor sample, and over half of the licensee sample. In both cases, no intentional emphasis was placed on any one industrial sector, but the parachemical, electrical and electronic industries do occupy a position which is relatively larger than the others. Lastly, the transfers of technologies we studied are more or less equally divided between North–North operations (between industrially advanced nations) and North–South operations (from a developed country to a less developed one).

The survey was based on questionnaires of a relatively closed nature administered by an interviewer; this limits difficulties of comprehension and certainly improves the quality of the data obtained. Obviously, this still depends largely on the reliability of the licensee's statements, for it seemed to us impossible to envisage using a direct approach method! Indeed it would have been utterly unrealistic for us to expect to collect data which very often were not available prior to the survey and had to be produced for this purpose. This is particularly true of the opportunity cost of technology transfers. This difficulty is common to all research on technology pricing, and indeed to the greater part of management research. This is, as it were, the 'rule of the game'.

One way to reduce the bias was to question people who had a good knowledge of the operation, because they had taken part in the negotiation, and in the everyday working of the agreement. This involved a variety of functions, depending on the size of the company: licence manager, export or marketing manager in the larger companies, general manager or external consultant in charge of the operation in the smaller ones.

The unit on which the survey was based was the licensing agreement, because we worked on the assumption that the licensor does not have one single strategy for all his contracts, but defines it according to the technology and the market concerned. This explains the fact that some interviewees completed two questionnaires.

The analysis was carried out on the whole duration of the contract, because the aim was to evaluate the price being paid, and payments are usually spread out over the period covered by the agreement. In most cases, we studied operations nearing completion, so that our interviewees were in a good position to evaluate the sums being paid. For a few contracts, the relationship had already come to an end, while for others it was just starting. But in all cases the agreement had effectively been signed. Generally speaking, most of the operations which were analysed had been running over five years (two-thirds of the licensor sample, four-fifths of the licensee sample).

For each of the contracts, questions were asked concerning the company's *objectives*, and the payment for technologies (all sums paid for the transfer of technologies). Other questions aimed at collecting data on a number of variables which can have an influence on the strategy being followed, or on pricing. This enabled us, among other things, to carry out further tests on the factors which determine prices, as defined by our predecessors.

Our study will be developed in three stages, which correspond to the three parts of this book. In Part I, entitled 'Analysis of Technology Pricing', we will examine the *theoretical bases* for technology pricing, and the *conclusions* of the limited volume of applied research which has been devoted to it. This critical analysis aims not only to show the inadequacy of these conclusions, but also to set forth the criteria necessary to redefine the terms of the problem – which includes, among other things, taking into account the licensing strategy.

Part II, 'Strategic Approach to Licensing', will attempt precisely to lay the ground for the licensing strategy to be taken into account. It will consist in *situating the licensing decision* in relation to the other factors of the company's internationalisation, in order to identify the strategic objectives of the licensor. We will also examine the strategies for acquiring technologies before we can build up a concept of the licensor–licensee relationship which discards the traditional view of the pseudo-market for technologies leading to a strategic approach of pricing.

Part III, 'Pricing Policies for Licensing Strategies' first introduces factors for an empirical validation of our approach, based on the surveys we have carried out, and then goes on to define prescriptive models for the benefit of the agents of technology transfers.

Part I
Analysis of Technology Pricing

In Part I, we will analyse the theoretical basis of technology pricing, then the applied research on the subject. It will be useful however, as a preliminary, to devote a chapter to further details on the conditions of remuneration of licensing agreements and their significance, in order to explain the notions of pricing.

Then we will study (Chapter 3) the factors determining the value of a certain technology, as it is identified by the neoclassical market theory referred to by most authors – not only research scholars, but also consultants and even certain agents. We will endeavour, at this point, to explain the hypotheses of this theory in order to understand the 'deviations' which can occur in its application.

Lastly, we will present (Chapter 4) the two studies based on empirical research which are today considered as authorities in the area of pricing policy for licensing agreements. They confirm certain theoretical principles, and challenge others, which makes their contribution important. But our critical review of these studies will demonstrate that they do not take sufficient account of the strategic dimension of licensing relationships.

2 Types of Remuneration and Forms of Payment in International Technology Transfers

The transfer of technologies entails a vast number of remunerations of various types, which in principle have very precise functions that are not always fully understood or expressed by the negotiators responsible for defining them.

Hence it is impossible, when reading a licensing contract, to identify the price which the parties have agreed on. One has to recalculate it using a method which follows the principles as far as possible, but which also takes the partners' logic into account. In practice, we will show that the payments made have to be considered as a whole, and then manipulated in various ways in order to come as close as possible to the real situation.

In order to understand and justify the price variables we will be using, we therefore have to go into some detail, and explain the reasons which lead us to select a specific concept of pricing, which in fact has been adopted by most authors. We will first examine the different types of remuneration and their theoretical functions. Then we will explain the definition we have come to by taking into account the actual conditions of payment.

2.1 THE DIFFERENT TYPES OF REMUNERATION AND THEIR FUNCTIONS

As a rule, among all the payments received by the licensor, there are only a few whose function it is to compensate for the transfer of technical information and the right to make use of it. This is due to the fact that a licensing contract usually covers several parallel and complementary services. Thus it is possible to distinguish, along with the sale of technologies (i.e., communication and authorisation for use) the services preceding the application of the contract on the one hand,

and the services which have to be provided when the technology is
made available to the licensee on the other. These two operations
generate costs or losses (including losses of information) which logically have to be compensated for; hence the existence of particular
forms of remuneration. But theoretically they have no influence on the
price paid for the technology, according to Hubert A. Janisewski and
Marc Besso[1] from whose work is drawn the greater part of the
remuneration survey which follows.[2]

Remuneration for Technology

Payments which are intended to remunerate the technology being
transferred can take two forms: they can be variable, or they can be
expressed as a fixed amount. In the first case, the licensee will pay
proportionately, according to a scale agreed on in advance; this is a
royalty. In the second case, he will pay a lump sum which has been
fixed as a result of negotiations. The choice between the two is
essentially independent of the level of remuneration, which should be
the same whichever system is used; the lump sum should correspond to
the aggregate discounted royalties.

We will review the various types of royalties and lump-sum payments
pointing out in what circumstances they are used and the way they are
interpreted by each of the two partners. We will also mention a
particular method of remuneration, close to payment in kind: this is
cross-licensing.

Running Royalties

This is the most common, and therefore the best-known, form which
can be defined as a payment calculated periodically (annually or
quarterly) on the basis of a use or performance indicator.

The royalty is determined as the product of a rate and a basis. The
rate is most usually a percentage and the basis is a measurable unit,
such as sales turnover or added value. Sometimes, especially when the
sale of the licensed product cannot be isolated, 'value rates' expressed
in monetary terms are used, which are multiplied by the number of
units produced or consumed.

Generally speaking, the use of royalties results in the supplier firm
taking on most of the risk, because its remuneration depends on the
use, and on the possible success, achieved by the licensee. In case of

failure, the supplier firm will be the loser, and conversely in case of success, its income can exceed its target. Consequently, it will accept remuneration in the form of running royalties only if it is not put off by the risk involved, and also if it considers the capability of the recipient firm to be adequate.

On the other hand, the licensee will usually tend to prefer this type of payment which, firstly, means that payments are spread over a period of time and, moreover, related to the success of the licence. One particular circumstance, however, could discourage the licensee: in the case of too great a commercial success, such large sums would have to be paid back that they would represent an excessive discounted cost. Also, with this type of payment, the licensee must submit to his partner checking all his figures to calculate the royalties.

These different attitudes on the part of licensors and licensees have led to the setting up of systems which overcome certain disadvantages: variable rate royalties, maximum and minimum levels of royalties, to name but a few. These various measures are naturally the object of intricate negotiations as their respective advantages are complementary and can be weighed against each other. But these arrangements can only partially correct the disadvantages inherent in this type of payment, its uncertainty and the need for checking. Hence the other type: the lump-sum payment.

The Lump-sum Payment

This is based on the principle, as its name implies, of fixing in advance the amount of the remuneration to be paid in one or more instalments, usually at the start of the contract.

It should be noted that this form of payment concerns essentially the sale of patent rights, which is a permanent transfer of industrial property, whereas royalties are applied to the transfers of licences which are, by nature, temporary.

However, the two parties to a licensing agreement may well agree to fix the price of a technology at the negotiation stage. The advantage for the transferor is clear: he is certain of receiving an income – although it will, however, be limited. The recipient, of course, carries the greater part of the risk, as he pays before the technology has proved to be successful; on the other hand, he will not have to suffer from his activity being scrutinised by his partner. This is frequently a determining factor in Eastern bloc countries.

Royalties and lump-sum payments are both intended to remunerate the same thing: the transfer of the technology. They are alternative forms of payment for the same remuneration. Alternative, without being necessarily exclusive, in so far as the parties can agree on a mixed system of remuneration: a lump sum completed by running royalties. This is sometimes referred to as a 'cash and royalties' arrangement, but the expression is somewhat ambiguous, because other 'cash' payments usually take place in a licensing agreement. In any case, the price will theoretically be the discounted sum of the two forms of payment.

Cross-licensing

A last type of payment is sometimes to be found: the exchange of licences or patents which are considered to be equivalent.

This consists in 'paying' for a technology with another technology, which obviously implies that the two parties consider that they have an 'equal value'. It also implies that the two companies have something to offer, which means that they have both reached the same level of technology. This type of remuneration is found in technology transfers between industrially advanced nations, especially in the pharmaceutical and electronic industries where research and development costs are considerable. Furthermore, some contracts stipulate that the licensee is under the obligation of communicating to his licensor free of charge the improvements he may have brought to the technology transferred.

But this type of payment, like the first two, applies to the remuneration of the technology itself. The forms of payment which we shall now turn to have theoretically nothing to do with the price of the technology under consideration.

Remunerations Prior to the Agreement

Licensing negotiations involve expenses or risks, which may be considerable. Those who have borne them try to share them with the other party. This takes the form of three specific types of remuneration.

Front-end Payment

A front-end payment is intended to cover expenses borne by the licensor prior to the signature of the agreement. These can be the costs of design work and technical literature, marketing surveys, technical demonstrations, negotiations, legal advice, etc. They are often more costly for the licensor than for the licensee.

It should be clearly understood that the remuneration does not concern the cost of the technology itself – i.e., the Research and Development expenses not yet written off. It relates only to the expenses incurred for the negotiations themselves. The cost of the technology should be covered, where appropriate, by the remunerations described earlier which are intended for this purpose. Although 'lump-sum payments' and 'front-end payments' are frequently confused, their functions are fundamentally different.

In practice, the parties may also consider that a front-end payment, serving to compensate the licensor, is nevertheless part of the marketing costs which should be borne by him in the last resort. In this case, the front-end payment will be wholly or partly deducted from future royalties. At least the licensor is certain to recover his costs, if the cooperation does not materialise.

Disclosure Fees

Before the licensing agreement is even signed, it has usually been necessary to provide the potential licensee with a minimum amount of technical and commercial information, in order to convince him of the benefit to be gained from the transaction. This obviously entails the risk that the prospect will not follow up, but will start to apply the technology using the information which has been provided. The risk will vary according to the technology considered, and its legal protection; but it is considerable for some categories of 'know-how' which are protected only by secrecy and are easily reproducible after observation.

The licensor can then claim 'disclosure fees' before proceeding with the demonstration of his technology. This remuneration is intended to determine how genuine the potential licensee's enquiry is, to encourage him to pursue the negotiation, and to compensate the licensor for any possible loss in case the operation did not result in a contract.

On the other hand, if the two partners decide at the end of the 'disclosures' to enter into an agreement, the fees already paid may be deducted from the royalties.

Option Fees

Option fees may be paid when the potential recipient wishes to obtain exclusive rights, without initially committing himself through a contract, because he wants to obtain additional commercial or technical information before taking a final decision. This remuneration may be demanded by the licensor in exchange for delaying the sale of his technology for a few months.

These fees are normally retained by the licensor if the contract is not signed at the end of the option period. On the other hand, they may be deducted from the remuneration for the technology if an agreement is concluded.

The common feature of the three preceding remunerations is that they are paid *before* the agreement is signed. They are frequently deductible from subsequent royalties, but this arrangement should not conceal their effective function which is to cover expenses, or a risk borne beforehand by the licensor.

Remuneration for Complementary Services

Technology transfers frequently – and increasingly often it seems – involve the supply of services for setting up a process and instructing the licensee's staff in its application. These services are carried out by the licensor, or subcontracted to firms specialised in engineering, technical assistance or training. They naturally entail expenses (salaries, travel and hotel expenses for technicians and engineers, etc.) which are payable by the recipient.

Engineering Fees

The licensor usually supplies, as part of his services, the engineering skills specific to his technology; when the recipient possesses equivalent skills, that is sufficient. But in the case of a recipient who is not so technically advanced as his supplier, the specific engineering skills have to be provided with all the technical information necessary for production, including information which may well be public property. When this service is provided by the licensor, which is unusual, it is called 'integrated engineering', but in most cases it is left to specialist consultants or service companies.

The basis for remunerating these services is far more objective than for the previous remunerations, since it is the cost. This is known as a 'cost plus' arrangement. In fact, rates are published by professional organisations, according to the experts' levels of qualifications, so that the licensee is himself capable of evaluating the approximate price to be paid.

Technical Assistance Fees

Sometimes – especially when the licensee does not have the required level of technical skill in the area – it becomes necessary for the licensor to provide help in the application of the technologies. This is no longer a question of communicating information or setting up the technologies, which is the function of engineering, but of providing assistance in the effective *use* of the technologies. This cannot normally be placed in the hands of another company, and remains the licensor's responsibility.

The rules for remuneration are broadly the same as for engineering services. But they may be applied in a less rigid manner because the licensor does not necessarily control this form of pricing. In any event, the principle of covering the costs remains valid (salary costs of technicians and engineers, mileage allowances, travel and hotel expenses, etc.) with a normal margin for the employer.

Consulting Fees

Lastly, consultants (financial, legal, marketing, etc.) may be called in. By definition, these are independent of the licensor, but their fees may be payable by one or other of the parties, or may be shared between them. The invoicing criteria are again fairly standard, as they are based on a daily rate which covers the margin and overhead costs of the company employing them, but does not include the expenses directly related to the consulting job (travel, hotels, technical literature, etc.). The budget for these services can nonetheless be estimated with sufficient accuracy, if only by inviting tenders.

As we can see, all remunerations for services complementary to a transfer of technologies follow the same principle of 'cost plus' which should make it possible to isolate them.

But a licensing contract usually entails several remunerations simultaneously. So one often sees agreements based on royalties, sometimes completed by an annual minimum, a front-end payment, and fees for technical assistance which is, as we know, the most common among the services (see Chapter 1). The calculation of the price of a technology implies that a distinction be made between the ancillary remunerations (initial and complementary), and those applicable to the technology itself. This should be facilitated by the fact that the systems for fixing the two types of remuneration follow rules which are basically different. Unfortunately, the application bears little resemblance to the principles, for several reasons which we will be examining in the second part of this chapter. This makes it necessary to forego the approach by individual remunerations, and to consider payment as a whole.

2.2 ACTUAL METHODS OF REMUNERATION: ANALYSIS AND APPRAISAL

We will first point out the disadvantages of the various systems of remunerations, which lead the parties to negotiate on a package, rather than on individual items. This is why we will recommend, in conclusion, that remuneration should be considered as a whole, by introducing the concepts of 'technology payments' and 'net technology remunerations'.

The Disadvantages of the Various Types of Remuneration

It should be made clear that negotiators often disregard the principles of remuneration which we have reviewed at the beginning of this chapter. They then agree on arrangements which make a detailed cost analysis impossible, and lead us to consider payments as a whole. Several reasons explain this behaviour.

The Complexity of the Principles of Remuneration

In practice, the apportionment of services is seldom coherent in respect of the principles described. One has to recognise, when questioning the people involved, that they do not always discern the subtle differences between the various types of remuneration. Hence the very common confusion, already referred to, between lump-sum and front-end payments.

In fact it is surprising to note that even the experts, consultants in technology transfers, voice contradictory opinions when justifying such and such remuneration. One could almost say that each consultant has his own definition, which only adds to the confusion. It is true that laying down principles is of limited interest for their practical application. How is it possible, for example, to make a practical evaluation of disclosure fees? How much should they amount to, in order to compensate a company adequately for the risk it is taking by disclosing its innovation during the initial contacts with a potential licensee? It is clear that subjective judgement plays an important part in this situation!

Conflicts Arising out of the Royalties System

Quite apart from possible disagreements on the level of prices to be paid, it should be pointed out that the principle of royalties in itself is a source of dispute between the licensor and the licensee. It will thus be in the licensor's interest to achieve maximum sales (in the case of royalties on sales), or maximum production (royalties on quantity produced), whereas the licensee's objective may be to maximise profit. But it can be demonstrated, as Manfred Perlitz has done, that these two objectives lead to different optima.[3] This means that, owing to the royalties system, the two parties will disagree on the sales objectives. It can therefore be expected that, for this additional reason, they will try to find compensations in order to avoid conflict in a situation where there are already many causes of disagreement.

Distortions Arising out of Differences in Inflation and Interest Rates

The fact that remunerations are usually paid over long periods of time makes them very sensitive to variations in inflation and interest rates. So an increase in the difference of inflation rates between the licensor's and the licensee's currencies is likely to modify the value of the remuneration paid over the duration of the contract.

Similarly, an increase in interest rates should lead to a rise in the discount rate which is used to calculate the expected remunerations, and should therefore result in their decrease. Contracts should allow for these distortions by including adjustment clauses in order to maintain the original value agreed on by the partners, as is done with other types of long-term contracts.[4] Such adjustment clauses are difficult to apply, and the parties here again have to agree on less complicated formulae.

These various reasons explain why remunerations are, in practice, fixed in a rather approximate way, so that if one of the parties feels he is losing on one item, he will try to make it up on another.

The Necessity of Evaluating Remuneration as a Whole

Compensations which may take place between the many types of remuneration are not due solely to the disadvantages we have just described.

Some licensors prefer to 'bundle' the various remunerations in order to negotiate a package without justifying the details of each item. This package reinforces the seller's negotiating position and enables him to dissimulate the inflated pricing of certain items.

Hence the necessity of evaluating the remuneration as a whole, by taking into account the aggregate of what has been paid during the contract. We will refer to this total amount as 'technology payments' which will therefore include not only remunerations relating to the technology itself, but also preliminary payments and those which are made as the technology is being set up for the licensee. It is clear that, by doing this, we will be somewhat overestimating the price of technologies, as we will be including payments which are not directly related to the technology being supplied.

A Definition of the Price of Technology

There is, however, one method of defining more precisely the real 'price' paid by the licensee. It consists in deducting the various costs arising out of the negotiation and the implementation of the transfer (the cost of the transfer) from the 'technology payments'. So the resulting sum will not include the remuneration of complementary services, nor negotiation expenses, and will provide a closer approximation of the price of the technology. We will define it as 'net remuneration of technology'.

We must point out that trading profits which may have been made by the licensor selling goods to, or buying goods from, the licensee are not included in technology payments, nor in the net amount after deduction of the cost of the transfer. This exclusion is necessary, precisely because we intend to analyse the influence of expected trading profits on the level of the technology remuneration.

We believe that adding these two types of income is a source of confusion. The amount paid by the licensee, which can be defined as 'the price of the technology being transferred', corresponds to the technology remuneration. This is the only payment he is committed to make by contract. Subsequently, the relationship can be interrupted.

On the other hand, the profit margins which the licensor can make on his exchanges with his partner are in no way guaranteed. They depend in fact essentially on the success of the relationship and on the partners' respective powers of negotiation. Of course these margins, if they are exceedingly high, do represent a cost for gaining access to the technology, as Constantine Vaitsos points out.[5] Nevertheless they cannot be considered as a 'price' in the economic sense, inasmuch as they are not a necessary condition for acquiring these technologies. It is clear that the transfer of technologies may well be followed by no commercial transaction between the partners – and, conversely, that these transactions may be pursued far beyond the period of technology transfers if the two parties continue to derive benefits from them.

This idea leads us to consider the factors determining the price of technologies which we will do, first from a theoretical standpoint and then by reporting on some applied research.

3 The Pure Theory of Technology Pricing

When one hears professionals justify the method they use to fix the price of their technologies, one is struck by their need to refer back to *principles* to show that their evaluations are not arbitrary. Reading *Les Nouvelles* (the review of the Licensing Executives' Society) is most interesting from this point of view. It contains many articles written by experts who endeavour to establish a connection between theory and practice.[1]

The 'determinants' of technology pricing put forward by professionals are frequently, though not exclusively, based on an economic theory of the transfer of proprietary rights. This is in fact the neoclassical microeconomic theory which, here again, has come to be regarded as the standard of reference. So it is impossible to discuss the usual factors determining the price without having previously discussed the theory itself; that will be the subject of this chapter. The aim here will not be to make an internal criticism of the model. More simply, we will examine the hypotheses which form the basis of the theoretical value of technology.

This critical analysis of what is hypothesised, implicitly or explicitly, will enable us to identify several determinants. Some will stem from the theory, while others on the contrary will come from the weakness of the hypotheses which form its base.

Before proceeding with a theoretical presentation (Section 3.2 below), we must first (Section 3.1) pose the premises of the theory, since any analysis of technology transfer pricing rests on the principle of appropriation in a system of industrial property. Then, in Section 3.3, we will examine the hypotheses of the model, and finally, in Section 3.4, identify the operational price determinants.

3.1 THE PREMISES OF THE THEORY: THE PRINCIPLE OF APPROPRIATION

In the ideal world of pure and perfect competition, technology does not have a price – or, rather, its price is nil, because free competition requires that technology be available free of charge. This condition means that all companies use the same technology, and that no company may enjoy a superior technology, as this would immediately be adopted at nil cost by its competitors.

This world does not exist. But it would be unfair to accuse its proponents of being unrealistic. The objective of this model is clearly not to describe a real situation; it is to show that free competition is the best principle for economic organisation because it enables the social optimum to be attained. Everything that could interfere with pure and perfect competition would impede the system, and should therefore be dismissed.

The Basis of Appropriation: Industrial Property

The system of industrial property is in itself a major distortion of the principles of free competition, because it results in the innovator's competitors being prevented from having access to the new technology. The patent limits control of a technology to its holder. He enjoys a *monopolistic position* which is contrary to the principles of the liberal economy. Most nations consider in fact that if they did not protect the interests of the inventor, there would be a proliferation of imitations and copies; the market would eliminate all profit, which would discourage innovation.

A decision of the Luxembourg Court of Justice expresses perfectly well this principle of granting appropriation to the innovator. 'On the question of patents, the specific object of industrial property is in particular to grant the holder, in order to reward the inventor for his creative effort, the exclusive right to use an invention for the manufacture and initial launching of industrial products, either directly, or by granting licences to third parties, and also the right to oppose any counterfeit copy'.[2]

So the state wants to allow the inventor to recoup his outlay – i.e., the expenses and risks borne by him in the research and development of the new technology. To achieve this, he is allowed to sell at a monopoly price, and thereby achieve a profit margin higher than the normal standard. Whether he makes use of the invention himself or grants a licence to a third party is irrelevant. He has on principle the right to retain a rent arising out of his invention – i.e., a profit exceeding the normal economic profit.

In fact, the granting of a patent – which means recognising a proprietary right on a technology – is not solely designed to encourage innovation. It also aims to promote the commercial use of innovations and the circulation of the technical knowledge related to the invention.[3] This is, in a way, a counterpart which the state expects from the beneficiary of its protection.

As far as encouraging the commercial use of inventions is concerned, the state also takes clients' interests into account, for they would otherwise be deprived of the benefits of technical progress. One can imagine that in the absence of patents, not only would there be less incentive to innovate, but the rare inventions would remain unexploited by their owners, who would see more risk than advantages. So the patent has the effect of reducing these risks and increasing the advantages of the application of a technical discovery.

Furthermore, the state wants to induce the innovator to whom it has given a proprietary right; to disclose the nature of his invention to the public. There are two reasons for this. First, the state wants to ensure, by disclosing it to all, including of course the experts in the particular field, that the technology really is new and therefore should be encouraged as an improvement. Secondly, the state wants to use its protective power to force the diffusion of information which might be useful to the progress of technical and scientific knowledge in society.

Thus appropriation is not based on economic factors, but on state intervention; it is constantly being challenged both by authors favourable to a completely liberal system (Fritz Machlup), and others who support an interventionist ideology (Constantine Vaitsos). But would doing away with industrial property make technologies unappropriable? Certainly not, because other factors intervene in the appropriation process.

Factors of Appropriation Unconnected with Industrial Property

Technology cannot be considered to be freely available even in the absence of proprietary rights. There are *market forces* which have the practical effect of giving control to the innovator. This applies not only to industrial secrecy which we mentioned earlier, but to a whole series of factors which have competitive consequences favourable to the innovator by allowing him a form of monopoly power.

F. M. Scherer[4] has listed three main causes which explain this situation:

- An imitation time lag.
- The competitive advantage of the innovator.
- Entry barriers.

Before an innovation can be imitated and effectively applied, there is inevitably a time lag. This lag is due in the first place to the fact that the innovator tries to keep his technical knowledge secret. Most countries recognise this in law. In fact, it is sometimes suggested that companies increasingly prefer this protection to patent rights.[5]

Furthermore, real markets have nothing in common with those of pure and perfect competition, where information is available immediately, and free of charge. In practice, time passes before some competitors realise the real benefit of the new product or process, and react by presenting a similar product. It is true that this 'incubation' period varies considerably from one industry to another, according to the complexity of the technology.

Lastly, the efficient application of a new technology does not require only knowledge and information. It usually entails the solution of practical problems where experience is the vital factor. Only experience allows the accumulation of know-how, that informal expertise which is necessary for the new product or process to be concretely mastered. But, we insist, the acquisition of know-how is costly. Even if the imitator tries to acquire it independently, he will have to spend time, and therefore money. Contrary to the implication of the industrial property theory – i.e., that in the absence of patents, new technologies would circulate freely and at no cost – the cost of imitation would not be negligible.[6] All this leads to the conclusion that the innovator will have a period of time available, variable according to his branch of activity, in which to apply his invention in a monopoly situation. It is not impossible, that in the meantime, he can pay off his Research and Development costs.

Beyond this period, competitors will generally be at a disadvantage compared to the leader who will enjoy a *competitive advantage*. Although this is not systematic, the initiator of a movement frequently enjoys a 'first mover advantage'. This is due first to the image effect, which will allow the initiator to differentiate himself from his followers, thereby enabling him to charge more lucrative prices and retain market share. Moreover his experience will enable him to 'move down' faster on the experience curve, and so enjoy lower production and marketing costs.

Lastly, when this innovation occurs in an industry which is already strongly oligopolistic, it is possible that major entry barriers will discourage potential imitators from entering the market. Only competitors already present in the industry will be in a position to engage in an imitation process, while outsiders will find it difficult to get together the resources, human skills and necessary distribution channels. In these circumstances, there is a strong likelihood that the industry will retain its oligopolistic character and imitations will not prevent the innovator from making large profits.

These considerations could be extended to the case of technologies which are not so new. We know that some technologies cannot be patented because they are based on well-known principles. Nonetheless, experience in the application of these technologies can give a competitive advantage to the company which has it. And this advantage may well be appropriate because no-one is capable of copying it without the help of this company. Finally, the actual property of a technology depends basically on the presence of a *market power*, whether this is accepted officially by the industrial property system, or recognised by competitive forces.[7]

This economic fact has been expressed in legal terms by certain lawyers who recognise the existence of proprietary rights for know-how by describing it as 'The body of knowledge for which a person wishing to make savings in time and money is prepared to pay a given amount of money'.[8] The recognition that the appropriation of a technology is a matter of fact, and not simply a matter of law could hardly be expressed more clearly. This in itself underlines the indissoluble link between appropriation and the price of a technology.

3.2 STATEMENT OF THE THEORY OF TECHNOLOGY PRICING: EXTRACTING THE RENT

The fact of the total or partial appropriation of technologies having been accepted, we can now turn to an analysis of pricing. The analysis is based on the principle that transferring a technology means giving up the advantages it produced – i.e., the income derived from the monopoly rent. The owner of the technology would sell only if he receives income equivalent to his 'opportunity cost' – i.e., the revenue he would derive from operating the technology himself.

This very broad principle, typically neoclassical as it does not take the conditions of production of the technology into account, underlies various analyses of diverse inspiration. It can be found among professionals like Orleans,[9] but also among academics like Caves and Murphy,[10] and with a supporter of the interests of developing countries like Vaitsos.[11] The last-named actually recommends to the governments of developing countries that they should determine the maximum price allowed for purchasing a technology according to the profitability it brings to the client, except in cases where no other possible suppliers are available. Such a measure has in fact been adopted as part of a resolution of the member states of the Andean Pact (Decision 24).

We will make an analysis of this principle in two stages – first static, then dynamic. Our objective here will be to determine the theoretical level of pricing in the model under consideration.

Static Analysis of the Rent

Two types of transfer of new technologies may be considered. They can be either a product, or a process. We will analyse them one after the other, but we will find few fundamental differences. In both cases, the prices will be equal to the monopoly rent allowed by the technology.

The Case of a New Product

Let us imagine a company which holds a patent on a new product. How will it determine the rent which will fix the theoretical selling price of the technology? According to the neoclassical model, it is sufficient to fulfil the conditions of optimisation of the monopoly firm, which means that the company has to produce and supply to the market such quantities that its marginal cost equals its marginal revenue.[12]

Graphically, maximum profit is represented by the surface π_1 in Figure 3.1.

So the owner of the new product will obtain maximum profit by supplying the market with Q_1, which, in the face of demand, will fix the market price at P_1. The surface π_1 is in fact what economists refer to as 'economic profit', meaning what remains after all the factors of production, including capital, have been remunerated. In financial terms it is super-profit, or a monopoly rent.

The holder of the patent will agree to transfer his property rights only if the buyer pays an amount *equivalent to this super-profit*. But it is most important to point out that the licensor (the seller of the technology) will have to leave the licensee a profit which will allow sufficient remuneration (i.e., at opportunity cost) for his factors of

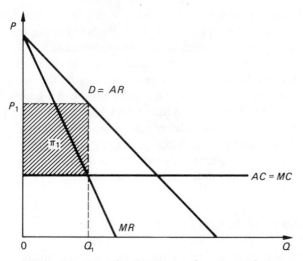

Figure 3.1   Rent for the owner of a new product
D is the demand curve, therefore the average revenue curve (*AR*)
AC is the average cost curve, taken as constant for simplification, therefore
equal to marginal cost (*MC*)
MR is the marginal revenue curve

production. The licensor will be able to retain only profits which are in
excess of normal profits. So the price of a technology is not determined,
in theory, by its total profitability – since the licensee would then have
no reason to enter into the agreement – but by that part of financial
profits which exceeds the normal rate.

The Case of a New Process

Let us imagine a company which owns a new process which enables it
to obtain production costs lower than those of its competitors. Its rent
will be determined by the same principle of optimisation, which means
that it will increase its production and sales up to the point where its
marginal cost equals the market price *P* which is the same for all
suppliers.

Figure 3.2 provides a graphic illustration of this situation. Maximum
profit is obtained with Q_2 and therefore amounts to π_2 (the shaded
area). While other companies on the market obtain no economic profit
over a long period (which means only a normal financial profit), the
owner of the new process will receive a super-profit. It is the value of
this super-profit, or rent, which will determine the selling price of the

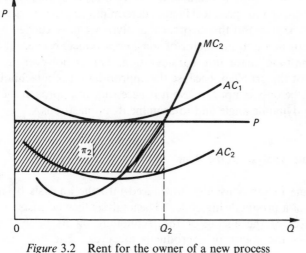

Figure 3.2   Rent for the owner of a new process
AC_1 is the competitors' average cost curve
AC_2 is the average cost curve of the owner of the process
MC_2 is the corresponding marginal cost curve

technology under consideration. So this is essentially similar to the previous case.

Now we must explain how the relationship between licensor and licensee will be set up. The seller will act as an auctioneer by selling the technology to the buyer offering the highest price. The bidders, who all have a perfect knowledge of the market, will each offer a purchase price, and the winner will be the one who has been most efficient in the application of the patent – or, more precisely, the one who will have anticipated the best application of it.[13]

This procedure is not unlike the tenders used for the concession of natural resources. In fact Bowler[14] reminds us that the term 'royalty' was used in the past to describe the king's share in the production of the mines for which he had granted a concession. The neoclassical theory of mining rent is also based on the bidding principle. Each candidate competing for the concession makes an estimate of the unit costs he is likely to obtain, compares them with the market price, and offers the owner a level of rent which will leave him sufficient profit – i.e., at least equal to a normal profit. If there is total competition between candidates, the rent should reach the optimum.[15]

This analogy does show that the theoretical price of a technology is not determined by the operating profit obtained from the industrial property being transferred, but only by that share of the profit which is in excess of a 'normal' remuneration. We must keep this in mind when we go through the practical factors determining a price, in the final part of this chapter. But the theoretical analysis we have carried out so far suffers from a limiting factor of some importance, because it does not take the time factor into consideration. This is however an essential aspect of the problem, because the appropriation of a technology can, at best, be only temporary. So it is necessary to approach the analysis from a dynamic angle and to examine its implications.

Dynamic Analysis of the Rent

The time factor is usually introduced into the analysis by using the model of a product's life cycle.[16] The analysis then consists in studying the effect of the ageing of the technology on its price. Figure 3.3 describes the stages in a product's life cycle. The first stage is the product's development, during which the company will invest in Research and Development (marked $D(t)$) prior to the perfecting of the new technology. Once the patent has been obtained, the product is

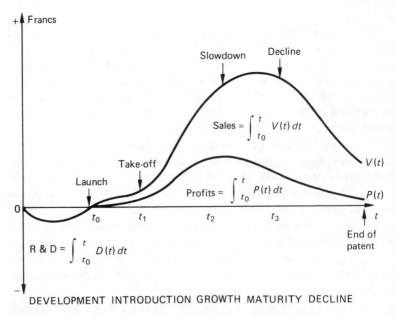

Figure 3.3　Life cycle of a patented product

launched on the market but its sales ($V(t)$ curve) grow slowly (distrust on the part of clients, defects discovered after using the product, etc.). Then the product takes off, and there is a period of growth, followed by a slowdown. The potential market has been reached; in the case of durable goods, the potential market has reached saturation point and the product enters a phase of maturity when sales will be practically static; this will be followed by a period of decline caused, among other things, by the arrival on the market of a substitute product providing the same service. The total of sales derived from the new technology is shown by the surface

$$\int_{t_0}^{t} V(t)dt$$

While going through this cycle, the holder of the patent will derive profits which will initially be negative (development period), then positive and increasing. His concern will be net profit as a whole – i.e., the difference between:

$$\int_{t_0}^{t} P(t)dt \text{ and } \int_{t_0}^{t} D(t)dt$$

In order to be coherent with the static model, we will assume that the profit is 'economic' – i.e., in excess of normal profit.

The price of the technology will be determined in the same way as previously, as the owner will accept to part with his 'possession' only if he receives a remuneration equivalent to his 'opportunity cost', that is, to the optimum profit he is giving up, which is represented by the surface

$$\int_{t_0}^{t} P(t)dt$$

Research and Development costs which have been incurred do not, in actual fact, have an effect on the price calculation, but they can be considered as a point of reference inasmuch as the licensor should not accept a remuneration below that level. So they form a kind of minimum price, below which the inventor does not 'cover his costs'. But it is the profit which may be derived from that technology which will determine, logically, to what extent this principle can be applied, as it is considered that the licensee will pay an equivalent price.

Another determinant of price appears in this dynamic model: the *time factor*. The value of a technology is a decreasing function of the time elapsed since it was introduced on the market. This is of course due to the reduction in the profit surface, as time elapses. This reduction is caused not only by a diminishing volume of sales, but by optimum pricing policy which an innovator has to follow, consisting in gradually reducing his selling price down to the minimum limit, in order to discourage potential imitators.[17]

Following this analysis of the theoretical approach to technology pricing, it may be interesting to recall the essential determinants of price it has brought to light:

- *Profitability*, and more precisely, the amount of 'super-profits' is the criterion for remuneration.
- This should *not be*, as a rule, *less than the expenses incurred* for developing the new technology.
- Lastly, it depends on the *time remaining* before the monopoly comes to an end.

But the two models under study rely on a number of implicit hypotheses of considerable weight. These must be analysed from a critical angle because they help to understand the difficulties in the practical use of the determinants we have just mentioned, and the necessity of finding somewhat approximate solutions, like the one we will present in the final part of this chapter.

3.3   THE IMPLICIT HYPOTHESES OF THE THEORY: AN IDEAL SITUATION FOR RENT

We will not cover, in this third section, all the simplifying hypotheses which the neoclassical microeconomic model relies on, as they have been largely reviewed elsewhere.[18] We will concentrate only on those hypotheses which are difficult to comply with in real life, of which there are essentially four.

First Hypothesis: the Licensor is in a Situation of Perfect Monopoly

This is a condition for the pricing policy to be valid in the static model (maximisation of profit in the face of a demand curve). The patent must therefore be perfect, in time and in its geographical dimensions, which means:

- That there is no other *competitive technology*.

– That the licensee is unable (technically and legally) to find a *similar technology*.

Now it is exremely rare, however original a given technology may be, that no alternative exists to supply the needs of the market.[19] This is due in the first place to technical reasons: solutions for supplying the same need may take completely different forms – e.g., cinematographic productions can be put on plastic film or on magnetic tape. So there is frequently some competition, at least marginal, which limits the inventor's degree of freedom. But the legal aspect is also important, because the holder of the patent has the right to claim the applications and the territories he is interested in. In fact, he has to define the boundaries of the monopoly he is requesting. If this definition is inadequate – possibly because something has been omitted – there is a risk that competition will appear. This is what happened in the case of the American inventor of the wind-surfing sailboard who forgot to take out a patent in France, the country which eventually became the largest market in the world. All this is even more applicable when the technology is not patented but is based on know-how, which, as we know, is of vital importance in contracts.

The first implication of the monopoly not being perfect is naturally that the rent will be lower than that computed by the model. But, more importantly, this situation implies that the licensor may not be the only person offering the licensees a technology. The sale of one or more substitute technologies will then be in competition, to a certain extent. According to the structure of the market, the licensees may be in a position to play off one licensor against another, in order for force them to bring down the prices, by giving up part of their rent.

Second Hypothesis: the Licensees are Competing for Access to the Technology

When the licensee agrees, in the theory under study, to pay the whole of the rent (or super-profit) which he himself draws from the market, it is clearly because he has no alternative, otherwise the licence would be granted to someone else.

But in fact the licensee may enjoy some market power – for instance, because he owns a trade mark which distinguishes him clearly from his competitors. The situation can even be completely opposite to that in the models if on the one hand substitute technologies are available, and on the other the licensee is in a monopolistic position in his market, which sometimes happens with public corporations in certain countries. This assumes, of course, that there is no restrictive agreement between the suppliers of technology. In any case, the drawing of the rent by the licensor can only be incomplete.

On this subject, we must mention one particular point. What happens if the licensee does not act to maximise profits but strives to attain other objectives, such as maximising sales? How can the rent to be shared be calculated? This, in principle, should not cause any difficulty, because the price of a technology is not determined by the rent effectively drawn by the licensee, but by that which the technology 'allows'. The fact that the licensee decides not to follow this principle of maximisation makes no difference: he will pay as if he did.

This remark brings up the formidable problem of the two partners' access to information.

Third Hypothesis: the Licensor and Licensee have a Perfect Knowledge of the Functions of Supply and Demand

For the licensee to determine the amount of rent he can pay, and for the licensor to accept his offer, they must have access to the same information on the quantities required, at different prices, and on production costs according to the quantities manufactured.

This is true not only for the short term, but also for the long term, as the period of appropriation can cover up to 20 years (the period of validity of a patent in most Western countries).

But the fixing of a price relies on information which is far from perfect, as it consists of forecasts which are always based on data which is questionable. The price of a technology is thus the result of a negotiation, rather than agreement on a calculation. Hence, in addition to the uncertainty felt by the two parties on the price calculation, there can be a doubt as to the validity of data supplied by the other party. It will indeed be in the licensor's interest to underestimate the costs and to overestimate the demand, whereas the licensee will be tempted to act in the opposite direction.

In conclusion, there is little chance for the amount of the rent to be known in advance, and for it to be drawn by the licensor from the licensee.

Fourth Hypothesis: the Licensor Behaves as a Seller of Technology

In the theoretical analysis which we have described, we make the implicit assumption that the owner of the technology is striving to draw the whole of the rent. Such behaviour would be acceptable if selling technology were the company's only (or even its main) activity. But very few licensors are in this situation, and in fact we have kept them outside the field of our investigation, which concerns only technological exchanges between industrial and commercial firms.

In the case of these companies, the transfer of their 'technological capital' is not carried out independently of their basic activity, which is to produce and sell products and/or services. From this point of view, it may be far more rewarding, as we will show in the third part of this book, to give up part of the rent in order to build up a commercial relationship with a partner, or even to lay the ground for a financial takeover.

This critical review of four determining hypotheses in the theory of technology pricing thus suggests that the rent will tend to be drawn only in part, which means that the licensor will be able to collect only a fraction of the super-profits derived from the technology. The size of this fraction – in other words, the price – will depend on the competitive environment of the two partners.

This result has considerable implications. In fact it cannot be asserted that one of the two partners will systematically have a dominant position in the agreement. It all depends in fact on the competitive environment in which the buyer and the seller operate. If the licensee is in a monopoly position, and the licensor has had to face very stiff competition in order to be selected, it is highly probable that the whole of the rent will remain in the hands of the buyer. The opposite applies if the licensee faces competition, while the licensor enjoys a perfect monopoly.

Hence it should not be considered as systematic or intrinsic to the system that the transfer of technologies will be unfair to the buyer. In some cases, it will take place to the licensor's advantage, in others to the licensee's, and in other cases still for the benefit of both. The sharing will depend on each partner's bargaining power, which is determined essentially by the relative intensity of competition within the licensors' and licensees' markets. So a degree of uncertainty can be expected; the shares of the super-profit drawn by the owner will vary from 0 to 1, according to the 'bargaining power' of the two partners.

This determining factor in technology pricing completes our review of the theoretical models. We can now place in perspective the determining factors used by professionals in their negotiations.

3.4   OPERATIONAL PRICE DETERMINANTS: THE ART OF APPROXIMATION

Most of the variables used by professionals originate, more or less directly, from the theory presented at the beginning of this chapter. But they are most often questionable extensions of it, so it may be worthwhile to attempt to evaluate their coherence and to link them with the theory on which they claim to be based.

There is no shortage of lists of factors to be taken into account to evaluate the price of a technology. We will only mention here those of Orleans,[20] Finnegan and Mintz,[21] Vaucher[22] and Mougeot.[23] Among all the factors they refer to, we will discuss those which appear in practically all the presentations. We have selected four. They are the profitability of a technology, Research and Development costs, opportunity costs, and lastly transfer costs.

The Profitability of a Technology

This first determinant is at the same time the one mentioned most frequently, and the one most closely related to the theory.

Taking this determinant into account requires a technical and economic study in several stages:

- Evaluate the *potential market* and the *market shares* which may be achieved over several years, in order to estimate the potential turnover.
- Estimate *total manufacturing costs*, so as to determine expected profits.
- Calculate, on these bases, the *Return on Investment* (ROI) which the licensee ought to achieve before paying royalties.
- *Share this profit* between licensor and licensee, and define the conditions of remuneration for the licensor's share (royalties, lump-sum payment, etc.).

This first method raises a number of practical and theoretical difficulties.

In practical terms, the results of the economic study will be so important that disagreements can be expected between partners on the validity of the data, especially if one considers the imprecise nature of market research. On the other hand, how is the share which the licensor will draw from the licensee's ROI to be determined? Reference standards have been suggested on this point. Finnegan and Mintz[24] refer to a commonly accepted yardstick of 25 per cent, which is supposedly justified by the fact that the licensee, taking on most of the investment and the risk, should receive the major share of the profit.

One could ask why this 'major share' comes to 75 per cent, and not 85 per cent or 60 per cent. But we will revert to this in detail in the next chapter, when we shall examine the statistical studies carried out by UNIDO on this question.

The Cost of Research and Development

The amount spent on Research and Development to develop the new technology is a variable frequently used by professionals.[25] It also appeared on the dynamic model which we have presented above.

Its use is justified by two fundamentally different considerations:

- On the part of the *licensor*, to know how much he is enabling his buyer to save in terms of costs, time and risks.
- On the part of the *licensee*, to estimate the maximum amount he is prepared to pay, above which he would find it more profitable to develop the technology himself.

It should be pointed out that these two arguments are consistent with the questioning of the first hypothesis. They involve a bargaining tactic based on the assumption that the licensee would be able to *reproduce* the technology. This therefore implies that the licensor's monopoly can be reduced, and that the appropriation is not complete.

But the validity of this determinant depends on the information available regarding the costs of developing the technology. These include not only expenses in terms of money (technical staff salaries, laboratory tests, etc.) but also the risks taken and the time required. In practice, evaluating this cost presents great difficulties which the majority of licensors cannot get round, for several reasons:

- They do not record these *expenses* in a separate heading in their accounts.[26] They are scattered among many types of expense account which shows that they are considered just as operating costs, and not as an investment.

- Few innovations are the result of an *identifiable Research and Development process*. In other words, it is most difficult to relate costs to one particular innovation, because this is only one of the results of a line of research. At best, the cost of a line of research may be known, but it is not possible to break it down among the various innovations which it has produced. The query on the notion of costs[27] comes up again.
- In the case of *know-how*, it is even more difficult to identify a cost, because this technical knowledge results from the experience of the methods and manufacturing departments, whose costs are included in the company's normal activities.

How then is the cost to be integrated when the price is being negotiated? Professionals usually suggest taking only part of it, on the one hand to give the licensee the impression he is making a big profit, on the other hand because buyers of technology often tend to underestimate the cost of perfecting the technology. Opinions differ as to the proportion which the licensor should retain, but we can quote Maurice Vaucher's yardstick: between one-third and a half.

Governments of countries which acquire technologies are increasingly taking Research and Development costs into account as justification for the value of a technology. In Argentina, the INTI (Instituto Nacional de Technologia Industrial), whose function it is to examine (and in some cases to authorise) technology transfers, has developed a method along these lines.[28] It consists in assuming that there must be a relationship between the level of royalties requested, and the Research and Development ratio of the company transferring the technology. When the ratio of Research and Development to sales is less than the rate of royalties included in the contract, the INTI requests a reduction in the royalties. This again raises the question of the proportion which should exist between these two variables; the solutions which have been suggested are lacking in theoretical foundation.

To conclude our discussion of this second determinant, it is interesting to note that its action is not necessarily consonant with the first. A technology may have been very costly to develop without being profitable (e.g., the Concorde supersonic aircraft) while the opposite can be true (e.g., the Pac-Man video game). What should the price be when the two determinants contradict one another? We will suggest a way of getting round this contradiction in the next chapter.

The Licensor's Opportunity Cost

Professionals often recommend that the licensor take his opportunity cost into account, and this attitude is, of course, totally in line with the theory. But while its definition does not raise any major problem, the practical evaluation results in very approximate figures.

The opportunity cost, as we know, is equivalent to the profit lost by not using a product in an optimum market. It is, in other words, and to use a more trivial expression, 'lost income'. So the licensor has to cover at least his 'lost income', otherwise it would be more profitable for him to keep his technology. Unfortunately, the evaluation of an opportunity cost is not simple. One would have to determine how much profit the best among the other applications of the technology would have brought the licensor. This research would require perfect information. In fact, even by limiting it to a few alternative applications, evaluating the profitability of a technology in a market will remain a fairly rough approximation. So professionals have to be content with a mere evaluation of whether or not 'something better could be done'.

In practice, the problem is very often solved by the opportunity cost being nil, because it is impossible for the licensor to apply his technology in any other way (only one potential licensee, unwillingness to invest, prohibition of foreign investments, etc.).

Transfer Costs

The theory of information brings to light another determinant: 'Transfer costs'. Indeed, for the licensor, the expected remuneration must be at least equal to the costs he incurs in making the transfer, without even considering the cost of the technology itself. The value of the message must exceed the cost of using the medium.

But, contrary to what the patent theory implies, the transfer of data does not take place at nil cost.[29] The reproduction of a technology at nil cost would be possible only if firms had perfect information at their disposal. Dismissing this hypothesis, it follows that the transfer of data between the two partners requires certain resources. It is even quite costly in many cases, as David Teece has pointed out.[30]

In a work which is an authority on this subject, David Teece evaluates and explains the level of transfer cost: 'Technology transfer costs are the costs of transmitting and absorbing the relevant firm-, system- and industry-specific knowledge to the extent that this is necessary for the effective transfer of the technology'.[31] He includes the following costs: cost of exchanging initial technical data (pre-engineering), engineering costs related to the transfer of the process or product design, Research and Development costs arising from the adaptation or modification of the technology, costs of the training prior to the start of manufacturing, and the operating losses during the launch period. The cost of manufacturing, transferring and installing equipment and tooling is not included in this definition. Transfer costs thus encompass those costs related to the transfer of 'unembodied' technology, in other words the 'software'. We will retain this definition for the remainder of our work.

In the transfers of technologies which he studied (29 operations) Teece shows that transfer cost are far from negligible, as they represent an average 20 per cent of total project costs, with, admittedly, fairly wide variations (from a minimum of 2.25 per cent to a maximum of 59 per cent). But it is important to note that about half the amount of transfer costs come from the last category – i.e., training expenses and operating losses during the take-off period. Expenses related to absorbing the technology seem to exceed the costs of the transfer itself.[32]

According to Teece, several factors explain the disparities noted in transfer costs between different operations:[33]

- The characteristics of the *supplier firm*, particularly its experience and expertise.
- The characteristics of the *technology* (complexity, degree of standardisation, etc.).
- The characteristics of the *recipient firm* (technical and managerial skills).
- The characteristics of the *recipient economy* (access to the necessary factors of production).

Be that as it may, the essential point we will draw from this research is the clear distinction between 'transfer costs' and 'technology costs'. If the latter term has to be considered as a possible determinant for the maximum price of the technology because it means an alternative cost for the licensee, this is not true of the former, which is more a minimum demanded by the licensor, as we indicated at the beginning of this section. However, transfer costs have characteristics which give them a distinctive place among all the determinants of technology pricing. In particular, they are the determinants which can be most accurately calculated, which is why they deserve special attention.

Other Factors

The four variables we have just presented are certainly the most important. Other formulas are to be found, less significant in our opinion, either because they are derived from the others, or because they are very imprecise.

Thus, for example, a good many professionals refer to 'industry standards' – i.e., standard royalty rates which are used in certain industrial sectors. In the pharmaceutical industry, for instance, the royalty rate varies between 10 and 15 per cent of net sales, whereas in the domestic appliance sector, it fluctuates around 2 per cent.[34] The reference to a sectoral royalty rate is based on the hypothesis that technologies being transferred in a particular industry yield comparable profitability, and/or require equivalent development costs. It would be easy to give many examples which disprove this. But these standards by industry, however unsatisfactory they may be, are often used by negotiators as a convenient reference.

In fact, the royalty rate is not a good basis of evaluation because it is only one form of technology payment among others, and also because it may lead to great disparities in financial payments according to the duration of the contract. But these industry standards should perhaps be considered more as an expedient chosen by the partners in order to reach agreement more rapidly, than as a contribution to pricing theory.

The penalty payable by a licensee infringing a patent should also be mentioned. From a strategic point of view, the buyer of a technology may weigh the price being requested against the penalty which he risks having to pay if he infringes a patent. So the seller must on the one hand take the strongest possible action against infringements to give the imitators the feeling that damages will be considerable and conviction probable. But on the other hand he must not fix too high a price so as not to encourage non-compliance with the patents he wishes to sell or grant. This argument is indirectly related to the level of appropriation of a technology, as counterfeiting is indeed a limit to industrial property. The same analysis leads to the conclusion that a technology has all the more value if it is well protected by patent, which means there are no or very few other ways to get round the patent.

The Specific Role of Each Determinant

In conclusion, we have to note the approximate character of most of the determinants we have examined – except, perhaps, as regards transfer costs. But more disturbing is the fact that each one can be justified. Consequently we need to relate them one with another in order to determine their respective roles in the fixing of prices.

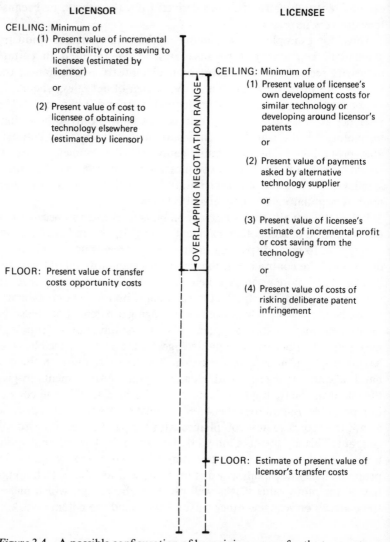

Figure 3.4 A possible configuration of bargaining ranges for the two partners
Source: Reprinted by permission of the publisher, from *International Technology Licensing* by Farok J. Contractor (Lexington, Mass.: Lexington Books, D. C. Heath and Company, Copyright 1981, D. C. Heath & Company).

This is one of the major contributions of Farok Contractor, the author of a remarkable book on remuneration in licensing agreements, which we will refer to at length in Chapter 4.[35] Contractor adds a further dimension, by showing the action of the determinants for each of the two partners – starting from the observation that, for each variable, the estimation will differ according to who is making it. Thus the licensee will usually underestimate costs and revenues compared with the licensor. One is then faced with two price ranges situated between a ceiling and a minimum. These limits are fixed by the various determinants (see Figure 3.4).

The juxtaposition of the two scales in Figure 3.4 produces a *negotiating range* within which agreement will be sought. This is situated between the licensee's ceiling, and the licensor's minimum. But although the initial divergence is thus reduced, the negotiating range usually remains very wide. After all, we are only describing a logical relationship, inspired by theoretical principles and practical constraints, between a body of determinants and the price of a technology. We have only a model which tells us nothing about the type of relationship or the relative weight of the various determinants.

These questions have been addressed essentially by two applied research projects which use widely different approaches:

- A *normative* approach, developed by a department of UNIDO, consisting in an attempt to evaluate the relationship between profitability and price, in order to lay down some rules. The basic idea is to concentrate on one of the hypothetical determinants and to evaluate its relationship to price so as to derive standards of remuneration.
- An *empirical* approach followed by the American academic Farok Contractor who has tried to identify statistically that (or those) determinant(s) which are most significant in the pricing of a technology. The emphasis is not laid here on a pre-selected variable, but on that (or those) which an empirical investigation has pinpointed as being the most significant.

Chapter 4 will be devoted to a presentation and criticisms of these decisive contributions to the question of remuneration in technology transfers. They are indeed a valuable source from which we can propose a new approach which will take their results into account, but will also strive to overcome their weaknesses.

4 Two Approaches to Pricing Policy

The two approaches we have chosen to study have some points in common, in spite of their methodological differences.

First, their explicit objective is to gain a better understanding of the action of price determinants, with a view to helping the decision making process. Secondly, they are both the most sophisticated versions that exist of certain theories on the determination of technology prices. From this point of view, they represent, if not trends of thinking, at least standpoints, which will also be the subject of our criticism. Lastly – and this point will hold a central position in this chapter – these two approaches imply that the licensor is considered as a firm whose activity consists in 'selling technologies'. Indeed, they suggest that the normal behaviour of the firm transferring a licence is to seek satisfactory compensation, a *'good' price*, without any consideration of the company's usual activity. This arises especially from the fact that the motives of the licensing strategy are never taken into account, which leads to the implicit conclusion that they are the same for all licensors, and that they can be boiled down to the simplest case, that where the sale of technology is an end in itself. This concept – which, in our opinion, is erroneous – is to be found in the normative approach as well as in the empirical one, as we will now show.

4.1 THE NORMATIVE APPROACH: SHARING THE PROFIT

As we have seen, the relationship between the profit derived from a technology and its price is directly inspired by microeconomic theory. But it has also been demonstrated empirically, as a study on franchising contracts has showed that the volume of fees paid by the franchisee was positively and significantly related to the net income he could derive from it.[1] Similarly, we know that the distribution of royalty rates between industries shows higher rates in those industries which are on average more profitable. Hence the idea that the price of a technology must be calculated as a share of the licensee's profit. So many recommendations have been made[2] concerning the sharing between

60

licensor and licensee; the percentage allocated to the licensor varies between 25 and 50 per cent, which is a fairly wide margin of uncertainty.

Some of the work carried out by the technology group of UNIDO Secretariat has consisted precisely in studying this point, by assembling as much data as possible on contracts, so as to identify the rules for sharing the profit.[3] This research is the source of the only data available today on the sharing of profit between licensor and licensee. We will present the method and results of UNIDO in our first part; then we will attempt to draw its implications, and finally we will point out the limitations of this normative approach.

The Relationship Between Royalties and the Licensee's Profit, and its Evaluation

UNIDO's evaluation of the share of the profit paid to the licensor calls upon several original variables which we must define before presenting the statistical results.

The UNIDO Model[4]

Let us suppose that royalties are based on sales net of taxes.[5] We can write the following equation:

$$\text{Sales royalty rate} = \frac{\text{Amount of royalties}}{\text{Net sales value}} \tag{4.1}$$

Now the 'amount of royalties' is a 'profit' for the licensor which, with this approach, derives from the profit obtained from the technology being transferred. Equation (4.1) can also be written as:

$$\text{Sales royalty rate} = \frac{\text{Licensor's profit}}{\text{Net sales value}} \tag{4.2}$$

Or:

$$\text{Sales royalty rate} = \frac{\text{Licensor's profit}}{\text{Licensee's profit}} \times \frac{\text{Licensee's profit}}{\text{Net sales value}} \tag{4.3}$$

The ratio

> Licensor's profit
> ――――――――――
> Licensee's profit

represents in fact the share which the licensor draws from the licensee's commercial profit, before payment of royalties and taxes. We will refer to it by the initials *LSEP* (Licensor's share of enterprise profit) which are used in UNIDO publications.

The ratio

> Licensee's Profit
> ――――――――――
> Net Sales Value

is simply the profit margin on sales.

We will refer to it as *POS* (Profit on sales) as UNIDO does. Later on, of course, the accounting definition of this term will have to be specified.

Lastly, we will show Sales royalty rate as *SRR*, with the same aim of being consistent with the UNIDO presentation.

So we have:

$$SRR = LSEP \times POS \qquad (4.4)$$

$$LSEP = \frac{SRR}{POS} \qquad (4.5)$$

The latter equation allows the share of the licensor in the licensee's profit to be determined, once the Sales royalty rate is known. Hence if the licensee makes a profit on sales (*POS*) of 20 per cent and pays royalties at the rate of 5 per cent, it follows that a quarter of his profits is taken by the licensor. However if, for some reason, his profit margin came down to 10 per cent, half of his profits would go to the licensor. Hence a contract must not be appreciated merely in terms of its royalty rate; the profit margin allowed by the technology must also be taken into consideration.

But the *LSEP* can also be calculated by the ratio using absolute values, with the amount of royalties paid as the numerator, and the amount of profit as denominator:

$$LSEP = \frac{\text{Amount of royalties}}{\text{Profit}}$$

However the 'profit' in the denominator usually cannot be isolated, as it would have to be the profit before tax and royalties, which the accounting system does not define. So it has to be recalculated by adding the royalties (R) to the profit before tax (PBT). The term 'royalty' is used here in the widest sense; it covers the whole of technology payments.

Thus we have:

$$LSEP = \frac{R}{PBT + R} \qquad (4.6)$$

By dividing the two terms of the relation by R, we get:

$$LSEP = \frac{1}{\dfrac{PBT + 1}{R}} \qquad (4.7)$$

in which

$$\frac{PBT}{R} = TTF \text{ (Technology turnover factor)}$$

Now the expression

$$\frac{PBT}{R}$$

has a most interesting significance because it indicates how many times the licensee has been able to recoup the royalties he has paid. Indeed if we consider that the licensee has achieved a PBT through the technology he has acquired, that he has had to pay R, then

$$\frac{PBT}{R}$$

is a kind of multiplier. The higher it is, the more efficient can the technology which he has acquired be considered, as it allows the licensee to recoup his costs several times over.

$$LSEP = \frac{1}{TTF + 1}$$

This second approach, based on absolute values, is particularly appropriate when the remuneration is not made up only of a royalty on sales (*SRR*), or when profit on sales (*POS*) varies from one year to the next. In these cases, which are in fact the most common, *LSEP* will have to be considered not year by year, but over the whole duration of the contract.

However, insofar as the calculation of *LSEP* is based on figures which apply to different years, we have to use discounted figures:

$$LSEP = \frac{\sum_{i=1}^{n} \dfrac{R_i}{(1+r)^i}}{\sum_{i=1}^{n} \dfrac{PBT_i + R_i}{(1+r)^i}}$$

where n = number of years in contract

r = discount factor

An example of this type of calculation is presented in Table 4.1 which follows. It does show that the share of the licensee's profit has to be considered over the whole period under study.

The Statistical Evaluation

Before presenting the evaluations which have been made of the variables in the sharing of the profit, it is necessary to provide some more precise accounting definitions of the values concerned.

1. Definition of the Variables   An initial distinction must be made according to the nature of the technology being transferred. Indeed all 'transfers of technology' do not lead to new sales for which the resulting profit would have to be estimated. In some cases, the technology being acquired does not generate sales, but results in a reduction in production costs. These savings are not always easy to estimate, but are clearly subject to sharing in such situations. So *PBT* can be either a profit which has been made, or savings which have been estimated.

Concerning the determination of profit, UNIDO recommends taking the difference between net sales (excluding taxes and freight) and total operating costs.[6] So it is a profit before tax and also before depreciation, which corresponds to the definition of 'profit before tax'.

Of course these definitions are applicable only if the required data is available. Otherwise one is 'forced' to work on estimations. Unfortu-

Table 4.1   An example of calculating the sharing of the profit according to the UNIDO method (units = dollars)

	Year 1	Year 2	Year 3	Year 4	Year 5
Net sales value	1200	1400	1800	2500	4000
Royalties (3% of net sales)	36	42	54	75	120
Profits before tax (*PBT*)	− 150	0	450	600	1300
Discount factor (rate 10%)	0.909	0.826	0.751	0.683	0.621
Present value of royalties (base year 0)	32.7	34.7	40.6	51.2	74.5
Present value of profits before tax (base year 0)	− 136.4	0	337.9	409.8	807.3

Sum of present value of royalties: 233.7
Sum of present value of profits before tax: 1418.6

$$LSEP = \frac{R}{PBT + R} = \frac{233.7}{1418.6 - 233.7} = 0.1414 = 14.14\%$$

Source: Arni (1984) p. 27.

nately, statistical studies using this method often seem to have been faced with this obstacle, which is one of their main weaknesses.

2. *Available Results*   In spite of their limitations, it seems to us interesting to present the results obtained by specialists from various countries using this method.

Five studies are available referring to three countries: the Philippines, India and Portugal. They have been carried out in a fairly similar way (see Table 4.2). Without giving extensive results, we simply show the overall figures provided by UNIDO.[7]

It is striking to note that the share drawn by the licensor in the licensee's profit turns out to be, on average, of the same order as that which is used by the experts (between 25 and 50 per cent).

But variations are considerable, as we see from the min. and max. columns in Table 4.2. The report from which we have drawn these results indicates that even in the case of contracts dealing with equivalent products, with similar terms and duration, considerable

Table 4.2   Evaluations of *LSEP* and *TTF*

Country	No. of contracts	LSEP (%) Arithmetical mean	Min.	Max.	TTF (%) Arithmetical mean
India 1	12	26.62	2.00	60.0	7.78
India 2	20	37.10	0.66	62.0	2.13
India 3	25	39.14	0.07	59.0	2.06
Portugal	14	33.27	12.20	70.6	3.09
Philippines	24	21.81	0.31	71.4	

variations can be noted. It should also be pointed out that, especially for the Portuguese sample, the method for evaluating the *LSEP* was not the same for all contracts.

The experts, Manuela Calixto Pires and Vitor Corado Simoes who carried out the study in Portugal, point out that they determined profit margins either on the basis of the companies' returns relating to their past profits, or using averages for each industry.[8] These limitations do not prevent the authors of the study listing the overall results to conclude that the search for a fair technological payment would lead to the conclusion that the *LSEP* rate should vary between 20 per cent and 50 per cent.[9] They do recognise, however, that this range should be more or less adapted to each country and industrial sector, to take account of the economic situation.

But, before coming to the criticism, we have to point out some of the implications of the study for the price analysis.

The Implications

It seems to us that four remarks should be made concerning these results:

- First of all, it appears that the average level of the levy on the licensee's profit can be considered as *relatively high*. A *LSEP* of the order of 30 per cent means in fact a royalty rate on profit of about one-third. It is understandable that such a levy is difficult to accept for a licensee, hence the practice of showing royalties as a percentage of sales, which has the added advantage of facilitating a check of the licensee's returns.

- The UNIDO method is also interesting because it brings into light a certain number of *factors which determine* the price of technologies, which are the justification of a high profit margin. This can be, for example, the novelty of the technology, the intensity of competition in the market, the size of the market covered by the licence.
- Thanks to these determinants, we have an explanation of the differences in royalty rates between industrial sectors; the most profitable sectors (such as the pharmaceutical industry) having rates which are *higher than other industries*. But differences which appear inside an industry are also understandable insofar as divergences in profitability compared to the average can be found in most industrial sectors. So UNIDO implicitly recommends that a technology's price should be fixed for each case individually, rather than by relying exclusively on sectoral standards.
- Ultimately, this approach results in the price of technologies being judged relatively. If this method is followed, the sum of money paid for acquiring a certain technology can logically vary from one contract to the other, owing to differences in profitability. The only possible yardstick for comparing these contracts is the *LSEP* ratio, which is an indicator of the relative price, as opposed to the absolute price which is the sum of the discounted value of royalties.

This research programme does therefore make a valuable contribution, particularly when compared to the vagueness which existed previously. Nonetheless, the considerable limitations of this method cannot be ignored.

The Limitations of UNIDO's Pragmatic Approach

The limitations of UNIDO's approach are due, in our opinion, to the theoretical and practical obstacles encountered.

The Theoretical Obstacles

Basically, UNIDO's method appears to be an application of the theoretical principle of linking the price of a technology to the profit it can generate.

But there is no real effort to validate this principle empirically. In particular, there is no examination of the form of the relationship between the sharing of the profit and the amount of profit. Yet there is nothing to prove that the *LSEP* is constant. In fact, according to the neoclassical theory, which UNIDO implicitly refers to, the *LSEP* should vary with the amount of profit, as the sharing should concern only the 'rent' – i.e., the profit in excess of the normal rate. So that if a technology brings only a 'normal' profit, *LSEP* should be nil, otherwise there would be no reason for the licensee to invest. Conversely, as the profitability increases, the *LSEP* can grow bigger, while still leaving the licensee with an adequate balance. This factor is apparently not taken into account by UNIDO, which merely points out the considerable variations in the *LSEP* in its samples (large standard deviations), without justifying them.

UNIDO is also faced with another difficulty, of the same origin, when it deals with the relationship between the royalty rate, and the share of the profit going to the licensor. Even if some studies appear to demonstrate it (Portugal and India 2), the *correlations are not significant* for the others (Philippines, India 1 and India 3).[10] Furthermore the two results which validate the relationship have opposite signs. We do not know if a higher level of *LSEP* leads to higher royalties (the case of Portugal) or lower royalties (the case of India)! Without going back to the problem of the quality of the data, it is clear that this question should not be treated without a more explicit theoretical framework, and without taking the whole of technology payments into account, and not only the royalty rate.

Lastly, all studies carried out using this approach ignore the part which the licensors' *strategies* may play in the fixing of prices, and therefore in the determination of the *LSEP*. But one cannot consider that all companies selling their technologies have similar objectives as far as remuneration is concerned. On the contrary, we believe that the share of the profit which is being claimed must depend, among other things, on the motives for selling the technologies. Hence some licensors will strive to maximise their *LSEP*, thus behaving in accordance with the theory, while others will be ready to give up part of their remuneration for the technology, in order to promote commercial transactions with their partner. Consequently it would not be surprising to find considerable variations in the *LSEP* from one contract to another. And a large part of the indeterminacy which is found could be explained by the diversity among the strategies being followed.

The Practical Obstacles

These theoretical obstacles are accompanied by considerable difficulties of practical application.

Indeed for the royalties to draw the share which has been agreed, the two agents of the transfer would have to know the expected profit at the outset. Forecasts are known to be unreliable. The best the partners can do is to agree on assumptions, but there is no guarantee that the actual results will agree with these assumptions.

Moreover, it often proves impossible to determine the profit which has actually been made, even at a later stage. In our own survey, 21 licensees out of 29 were unable to provide such estimations. This is due essentially to the absence, in most cases, of separate accounts for each licence contract. The technologies which are acquired are usually integrated in complex activities of which it is usually possible to evaluate only the contribution to the overall profit. It is exceptional for the object of the transfer to correspond to an activity which can be isolated from an accounting point of view. It follows that the calculation of the actual *LSEP*, which would be the only way to estimate its importance statistically, is virtually impossible, hence the approximations which have been made by researchers in this field.

Nonetheless, we believe that even if the UNIDO method comes up against practical obstacles for evaluating the *LSEP*, and so for identifying 'relative prices', it can be applied usefully for understanding pricing policies (i.e., intentions) by the analysis of licensing projects. It is highly probable that sales and profit will not correspond exactly with the estimates, but the expected royalties denote a remuneration policy – or rather, result – from the confrontation of the two partners' policies.

This last remark refers to a dimension which is hardly present in UNIDO's works, empirical validation. Their approach consisted in promoting a tool for evaluation, the *LSEP*, and in suggesting the existence of standards for technology pricing. This leaves open the question of the empirical identification of price determinants, a question which has been addressed by Farok Contractor, who has come to the conclusion that transfer costs play a leading part.

4.2 THE EMPIRICAL APPROACH: THE IMPORTANCE OF THE TRANSFER COST

In this second approach, the importance of the transfer cost is twofold:

- Important by its *value*, with the confirmation of Teece's conclusions that technologies are not transferred at nil cost,
- Important also by its *influence on the level of prices*, as the transfer cost is used by negotiators as a key variable they refer to in the negotiation of technology payments.

So it is a contribution which is very different in nature from the preceding one, at the same time complementary and contradictory, but in any event most valuable for improving the knowledge of technology pricing.

It is due to Farok J. Contractor who initiated this work at the 'Multinational Enterprise Unit' of Wharton School, University of Pennsylvania, under the direction of Franklin R. Root and Howard V. Perlmutter. The essence of the approach to the problem and the results obtained appears in a book by Contractor[11] but summary presentations are available in several articles.[12] We will present Contractor's work in the first part, then we will draw its implications, before formulating several criticisms concerning this research.

The Empirical Analysis of Price Determinants by F. Contractor

Contractor's method is certainly more in the tradition of management science than UNIDO's, since it consists in formulating hypotheses as regards price determinants, then in attempting to validate them with a sample.

Hypotheses were formulated using the negotiation range model introduced in Chapter 3. Contractor strove to identify all the factors intervening in the formulation of a price, and to express them in the form of variables which could be measured in the surveys. Hence he took into consideration, as assumptions, the part played by the age of the technology, the existence of a patent, the size of the production plant acquired by the licensee, etc. But above all, this model throws light on a particular variable which is the transfer cost, not only as a factor determining the minimum price, but also as a 'standard' for evaluating a licence by calculating the margin on the transfer cost (the net technology return) or the multiple of the transfer cost.

We will deal first with the method used, before presenting the main results.

The Method

In the first place, we need to define the price variables, as those used by Contractor differ significantly from those of the UNIDO method.

– He does not limit the 'absolute price' variables to royalties, or even only to contractual technology payments. He considers that the price paid is the *whole of the income* derived from the agreement. To the usual return, he adds the *profit margins* on transactions arising from the transfer, in the way of sales or purchases of products (see Table 4.3).
– He also advances a list of costs to be taken into account. These are: *transfer costs* which include four categories (see Table 4.3), R and D expenses and opportunity cost.
– The period used for measuring the 'net price' must be the whole duration of the contract, and not only one year, as there can be compensation over a period of time (a high initial payment resulting in reduced royalties, and vice-versa). This of course raises the problem of computing net present values.

In conclusion, Contractor expresses net return as:

$$
\begin{array}{l}
\text{Present value} \\
\text{of contribution} \\
\text{margin}
\end{array}
\quad = \quad
\sum_{i=1}^{n} \frac{\displaystyle\sum_{i=1}^{8} RET_{it} - \sum_{j=1}^{5} COST_{jt}}{(1 + r)^t}
$$

So we have at our disposal a genuinely original accounting framework, which we will in fact partly use in our study, with a few modifications arising from our criticism of this second approach.

On this basis, Contractor has attempted to find the determinants of the level of revenues (the 'absolute price') and of the contribution margins ('net price'). To achieve this, he proceeded in two stages:

– First he conducted in-depth semi-directive interviews with 12 senior executives on the parameters they took into account in determining a price.
– Then a questionnaire was mailed to 39 companies so as to collect data on their income, expenses and certain other items of information concerning themselves or their partners, which are supposed to have an influence on the price of the technology. Altogether 109 contracts signed by these 39 firms were analysed.

Table 4.3 Categories of returns and costs of technology-supplier firms over an agreement's life

Returns to supplier firm in year t

RET_{1t}: Front-end or lump-sum fees
RET_{2t}: Royalties
RET_{3t}: Technical-assistance fees
RET_{4t}: Fees for other specific services rendered
RET_{5t}: Payment in equity of recipient firm and dividends thereon
RET_{6t}: Net margins and commissions on materials or goods supplied or received
RET_{7t}: Value of grantbacks (improvements or innovations made by licensee)
RET_{8t}: Tax savings arising from arrangement, if any

Transfer costs incurred by supplier firm in effecting the agreement in year t

$COST_{1t}$: Technical services (direct and overhead)
$COST_{2t}$: Legal costs (direct and overhead)
$COST_{3t}$: Marketing assistance to recipient
$COST_{4t}$: Travel and management personnel costs (not included above)
$COST_{5t}$: Other direct costs associated with executing agreement

Other categories of costs

O_1: Total of sunk or development costs for the product or process transferred, up to inception of agreement
O_2: Opportunity costs (for example, losing export sales or direct investment opportunities in licensee's country or in other territories)

Source: F. Contractor (1981a) p. 35.

But it is important to point out several characteristics of this sample which concerns:

- Only American companies, of which three-quarters (31 out of 39) belong to the 500 largest corporations listed by *Fortune* magazine.
- Contracts which have been examined only from the sellers' point of view.
- Essentially transfers towards industrially advanced countries, which make up three-quarters of the sample, which is in fact close to what we know of the real situation (see Chapter 1).
- Only transfers of technology to independent licensees, because Contractor subscribes to the theory that internal transfers create bias in the research on price determination.

The Results

Contractor obtains a relatively large number of results, and it is out of the question to analyse all of them. We will merely refer to those which appear to us the most significant. First we will report on the conclusions arising from the semi-directive interviews, then we will present the statistical results.

1. Factors Taken into Consideration by Negotiators Interviewees were requested to classify various factors on a pre-selected list, according to the weight they attached to each one in the price negotiation.

Three of them stand out among the others (see Table 4.4):

- The amount of *technical and other services* provided to the licensee – in other words, the transfer cost, as this word expresses the idea in financial terms.
- *Industry norms* – i.e., the royalty rates usually applied in the industry. The inherent weakness of this price criterion is well known, but Contractor rightly points out that in the case of mature technologies, when competition between licensors is strong, this norm is in a way the alternative cost of the technology, so that it is difficult to disregard it.
- *The licensee's market size and profitability* appear almost as decisive as the norms in the negotiators' statements, so this seems to confirm the theory. But one wonders whether the answers on this point were influenced more by principle than by experience, since, on the other hand only a quarter of the interviewees mention 'the possibility or

Table 4.4   Considerations used by American executives in setting the price of a technology[1]

Rank	Criteria[2]	Score[3]
1	Amount of technical and other services provided to licensee	127
2	Industry norms	105
3	Licensee's market size and profitability	102
4	Take what's available	61
5	R&D expenditure	50
6	Returns must at least equal those from exporting or direct investment	28
7	Less for old or obsolescent technology	27
8	Less when patent expiring	10
9	Grantbacks	5
10	Patent coverage	3

Notes:
1. The sample was made up of 37 firms
2. Criteria taken from the negotiation range model and other criteria found in the literature on licensing were shown to executives in random order
3. Interviewees were requested to pick 5 criteria among the 10 giving a weight of 5 to the most important, 4 to the next one and so on. Criteria which were not chosen were marked 0.

Source: Root and Contractor (1981) p. 26

desirability of a quantitative estimation of the extra profit brought to the licensee'.[13]

So the statistical results on the economic and financial content of contracts do shed interesting light on these determinants.

2.   The Levels of Revenue and Cost Variables   Contractor's results on the level of *transfer costs* confirm, first of all, those obtained by Teece – i.e., a relatively high average level, usually above US$100 000 (see Table 4.5). This means, on the one hand, that the hypothesis of a nil marginal cost for the reproduction of technologies is once again proved wrong and that, for this reason, there are economic obstacles to imitations in addition to those arising from the system of industrial property. On the other hand, the fact that transfer costs are high involves a risk for the licensor who usually incurs these expenses before receiving any income. So this variable must intervene in the setting of the price, one way or another.

Table 4.5 Average total returns and transfer costs over the duration of agreements

| Category | Totals at current values | | | Present value with 15% discount factor | | |
	Cap.C[1]	Soc.C[1]	Dev.C[1]	Cap.C	Soc.C	Dev.C
Returns	3915	2066	3284	411	382	422
Costs	192	349	280	85	156	138
(Multiple)[2]	(45.94)	(12.7)	(9.14)	(34.96)	(13.65)	(8.04)
Contribution margin or rent	3720	1717	3004	326	226	284

Notes:
1. Cap.C: Capitalist countries
   Soc.C: Comecon and socialist countries
   Dev.C: Developing countries
2. The figure shown under (Multiple) gives the mean ratio of total returns to transfer costs for *each* agreement, which is not equal to the ratio of mean returns over mean costs in each column. The flow of income and costs are present values using a 15 per cent discount factor to allow for the effects of the time lag between income and costs over the duration of the agreement. For the complete sample, the means of the 'multiple' ratio for the series of figures at their gross and present values were respectively 35.67 and 27.63.

Source: Contractor (1981a) p. 105.

Parallel to this, revenues also prove to be considerable, exceeding 1 million dollars at current values. But more interesting is the comparison between total income and transfer cost. Two variables are suggested by Contractor for relating income to expenses. First, what he calls the *Multiple*, which is the ratio of income over expenses; this appears to be very high, 35.67 on average and 27.63 if the series is updated. A transfer of technology may thus be regarded as yielding a return of roughly 30 times its direct cost, which is a margin of 3000 per cent.

Of course, considerable variations in this multiple appear between one contract and another. More specifically, in 8 cases income was below transfer costs, which means that the licensor did not recoup his investment.

In the same way, impressive differences appear between economic zones. Contrary to what is often held, the contribution margin paid by developing countries turns out to be lower than that paid by industrially advanced nations. But this does not allow us to conclude that the price of technologies is lower for developing countries, because the multiple ratio is not in itself a perfect standard of comparison.

3. The Statistically Determining Variables   Among the various price variables used, the amounts of royalties and total income could be explained statistically (R^2 sufficient). As opposed to this, the regression equations brought no significant result concerning the contribution margin or the multiple ratio. Hence the variables defined relate to the notion of 'absolute price', one in fact being a component of the other (royalties in relation to total income).

Now if we examine the determining variables (those with one or two asterisks (*) in Table 4.6), it is interesting to note the strong presence of the *transfer cost*. First with the C_1 variable, but also with C_2 (technical costs) which make up the greater part of transfer costs, and lastly with S (scale of licensee's production plant) insofar as a large scale can result in higher costs.

So we have a statistical *confirmation* of the interviews of the executives, since the same criterion comes out as the most determining. As for the other variables, it is still possible to conclude, obviously with caution, that some of them appear to have a positive effect ($+$ sign) on the level of prices (the licence's life span, the transfer of a trademark, export licences), while others would seem to have a negative effect (duration of the agreement, degree of competition between the technologies' potential suppliers).

The Implications

Several lessons can be drawn from Contractor's analysis for practical purposes.

First of all, of course, the part played by the transfer cost in setting the price. Not only is a statistical correlation brought to light, but in addition this variable is deliberately taken into consideration by negotiators. This *validates*, at least in part, the model suggested by Root and Contractor which we presented in Chapter 3.

Furthermore, the results of this research show that the transfer of a technology is a *'profitable' operation*, since contribution margins are considerable, on average. But it should be borne in mind that this calculation of profitability is partial, inasmuch as on one side it takes direct and indirect income into account, while on the other side only

Table 4.6   Essential results of regression calculations carried out by F. Contractor

Equations selected

R_2 = $-0.195 + 0.297\ C_1{}^{**} - 0.191\ C_2{}^{*} - 0.236\ L + 0.349\ P + 0.257\ K$
$+ 0.622\ E - 0.194\ D + 0.135\ S^{**} - 0.175\ T_1 + 0.301\ T_2 - 0.759\ T_3$

Degrees of freedom = (11.36) $R^2 = 0.63$ $\bar{R}^2 = 0.52$
F-value = 5.894

R_3 = $-0.142 + 0.325\ C_1{}^{**} - 0.216\ C_2{}^{*} - 0.241\ L + 0.368\ P + 0.341\ K$
$- 0.766\ D + 0.125\ S^{**} - 0.109\ T_1 + 0.324\ T_2 - 0.138\ T_3$

Degrees of freedom = (10.39) $R^2 = 0.66$ $\bar{R}^2 = 0.57$
F-value = 7.415

**	= significant at 1% level or better
*	= significant at 5% level or better

Variables

R_2	=	Total royalties
R_3	=	Total returns
C_1	=	Total transfer costs
C_2	=	Total technical costs
L	=	Agreement life
P	=	Remaining patent life
K	=	Trademark
E	=	Export licence
D	=	R&D expenditure by supplier firm in 1976
S	=	Recipient's plant scale as a ratio of supplier's typical plant
T_1	=	Degree of supplier competition
T_2	=	Adaptation of technology for recipient

Source: Root and Contractor (1981) p. 32.

direct costs are taken into consideration. Further evidence on this point can be found in Bidault and Mullor.[14]

Lastly, licensors probably do not behave with a view to maximising profit, but seek rather to obtain a satisfactory remuneration, as Root and Contractor point out. In fact, negotiators do not usually carry out studies of the licensee's opportunity cost or expected profits, but usually rely on fairly rough estimations of the value of the technology transferred. This is in broad agreement with the bounded rationality theory (March and Simon).[15] But this attitude is also claimed by the licensors to be necessary. If the rent was drawn in full, they say, it would be to the detriment of a good climate of cooperation which is essential to the success of the operation, and the licensee would be tempted to 'cheat' so as to obtain higher profits.

These three conclusions are unquestionable contributions to research on technology pricing, but should not allow us to forget the considerable limitations of this approach.

The Limitations of Contractor's Empirical Analysis

Among the limitations which can be mentioned, some concern Contractor's methodology, while others relate to his understanding of the problem. But before examining them, we must point out the indeterminacy which this research leads to.

It is indeed regrettable that the determination of the 'Multiple' cannot be explained, as Contractor himself admits. We discover that total returns are a considerable multiple of costs, but this ratio varies from one contract to another, without any explanation being given as to the reason. This is obviously a considerable limitation from a practical point of view; negotiators cannot find in Contractor's conclusions recommendations as to the fixing of a price in relation to the transfer cost. So we will attempt, in the third part of this book, to reduce somewhat the indeterminacy which is due, in our view, to weaknesses of approach and methodology.

Methodological Weaknesses

The results obtained obviously depend upon the method used. Now this is questionable, for several reasons.

The sample, to start with, contains large corporations, essentially American multinationals, whose pricing policy has no reason to be similar to that of firms in other countries, and of smaller size. Furthermore, only licensors have been questioned, which probably creates a second bias inasmuch as these firms have certain types of information in their possession, but have no access to others. It is therefore necessary to use a more broadly-based and more international sample, as James D. Goodnow recommends.[16] This is what we have endeavoured to do, by questioning not only French licensors, but also French, Moroccan and Portuguese licensees, of various sizes.

A second weakness stems from the *absence of certain variables* in the statistical results. For example, the licensee's market, or his profitability, are not taken into account in the regressions, whereas the negotiators laid great emphasis on these variables which also hold a key position in the theory. Regression calculations obviously can evaluate only those variables which have been introduced, and our criticism regarding Contractor is precisely that he did not include these variables in his analysis. In fact, this is probably explained by the structure of his sample, which excludes licensees.

A final methodological weakness we wish to point out was failing to make the distinction between *technology payments per se* and the *profit margins* arising from commercial transactions between the partners. Now adding the latter to the former is interesting only when trying to establish a cost–advantage comparison for technology transfers.[17] But it seems to us erroneous to include remunerations which relate to a different aspect when doing research on technology pricing. In fact, these two components of total income can lead to *decisions* it would be interesting to study – e.g., if the objective of increasing profit results in a reduction of technology payments. This last point leads us directly to our objection in respect of Contractor's approach to the problem.

The Weaknesses of the Approach

Among the factors determining the price which were selected at the outset, there is no element concerning the strategies followed by the licensors. Contractor implicitly assumes that licensing is a strategic operation with the same ultimate objectives for all companies. But one can imagine, as we will show in the second part of this book, that various motivations can result in the transfer of a technology and that, according to the motivation, different pricing policies will be adopted.

Hence one wonders whether the eight contracts which proved unprofitable are the consequence of a failure in carrying out the transfer, or on the contrary are a result which was deliberately sought, for example to gain other advantages. The history of technology transfers has witnessed a good many such operations. But by excluding licensing strategies, Contractor deprives himself of one of the keys for explaining the relationship between income and transfer costs.

Ultimately the two principal approaches to pricing policy stumble over the *same obstacle* – i.e., the strategic nature of a licence. So it seems essential to include the *licensing strategy* as a factor of pricing policy, which is why we will devote the following part to licensing in a firm's strategy. We can then revert to the problem of pricing, in the third and final part, being then better equipped. But of course we will not neglect the important lessons learned from the theory, as much as from the research carried out by UNIDO and Contractor.

We already intend to select a few factors especially with a view to

building models for calculating technology prices; some variables do
appear to play an essential part. We will bear them in mind when we go
into the strategy for an optimum valorisation of the technology; this
case cannot be generalised in our opinion, but it does correspond to a
real situation, as we will demonstrate.

Part II
Strategic Approach to Licensing

Part II is essential to our analysis of technology pricing, since it allows us to go beyond the limits of previous works.

Our objective is to emphasise the strategic aspects of technology transfers, not only from the point of view of the supplier (licensor), but also from that of the recipient (licensee). We aim to do nothing less than call into question the traditional view of technology licensing, where the motivation is merely financial, since the licensor is supposed to be interested only in turning his knowledge to profit. We consider that this notion succeeds in creating an *artificial parallel* between technology and goods, and thereby dealing with technology pricing on the basis of false premises.

We will demonstrate that, on the contrary, licensing does have a strategic aspect insofar as it can have a lasting effect on the growth of a firm. Once the actors of this 'technology transfer' have assimilated this aspect, it is no longer possible to consider the price as a mere compensation. It greatly depends on the strategic motivations the two parties may have in mind for getting into contact.

We will begin by showing how technology transfer is *motivated in different ways* by introducing a taxonomy of generic licensing strategies (Chapter 5). We will take this opportunity to present some elements to validate this analysis, based on the results of our surveys. We will see that our taxonomy of motivations will play an important part when we come back to discussing pricing matters.

We will then study (Chapter 6) the *licensee's motivations*, in which specialists have not, until now, been really very interested. We will endeavour to fill the gap by stressing that, here again, technology acquisition strategies correspond to various motivations which we will identify with the help of previous research.

Finally, we will be able to postulate a theory of the way technology sales and purchase strategies fit together in a theoretical approach to the *licensor–licensee relationship* (Chapter 7). We will go beyond a theoretical view which is exclusively economic and integrate institutional and strategic elements in order to demonstrate the interactions between both partners to a licensing agreement. This model will constitute the theoretical basis of the pricing analysis discussed in the third part.

5 Licensing Out Strategies

The strategic aspect of licensing out has generally been neglected. It is often considered as a mere alternative, be it an alternative to other forms of internationalisation or other forms of technological development. In fact, one simply presents a long catalogue of circumstances where licensing proves appropriate.

But we consider it necessary to go beyond this approach and build an *analytical model* of the *motivations and means* used in licensing out technology. Such is the first aim of this chapter, in which we present the generic strategies of licensing out. We shall see that in spite of a large number of different situations, technology transfers correspond to *three possible strategies*.

After presenting these strategies and appraising their relative frequency, we shall show how they enable us to explain the inclusion of licensing in a firm's corporate strategy. Indeed, if we want to go beyond the narrow concept of technology transfer as an 'opportunistic' action, it is necessary to give an accurate picture of the position occupied by licensing strategies within the firm's 'overall project'.

5.1 GENERIC STRATEGIES FOR LICENSING OUT TECHNOLOGY

Working out a taxonomy of licensing strategies entails giving ourselves classification criteria for licensing out agreements. It is clear, however, that the choice of these criteria is decisive, for many different classifications can be considered. Qualifying these classifications as 'strategic' means that they should be based simultaneously on the objectives and the means used to reach them.

Although the term 'strategy' has often been used incorrectly, there is now a general consensus to consider it to mean steps taken to determine 'objectives, purposes, or goals, ... and the principal policies and plans for achieving those goals'.[1]

A generic strategy for licensing will thus designate a type of conclusions and motivations which will be expressed by the use of certain means, policies and practices between the partners. This study will lead us to distinguish three types of licensing out according to the *motiva-*

tions and *practices* they involve. This distinction seems more appropriate to us than the multiple distinctions which oppose so-called 'offensive' strategies to other so-called 'defensive' or yet again 'opportunistic' strategies insofar as they stress the firm's proactive or reactive attitude, but do not concern the 'stakes' of the action.

We shall therefore begin by presenting our taxonomy of licensing strategies and then give a few key elements in order to evaluate their respective importance.

Taxonomy of Technology Licensing Out Strategies

Taking into account the objectives and means of licensing out, we can distinguish three generic strategies:

- *Market strategy*, by which the firm licenses out its technology in order to open up a new market for its own products,
- *Production strategy*, by which the licensor seeks rather to improve the firm's supplies, costwise or qualitywise, for certain products, whether manufactured or not,
- *Technology strategy*, whereby the valorisation of appropriate technologies constitutes the licensor's only aim. The licensor subsequently nurses no other ambition in respect of his client, except of course receiving a satisfactory compensation from him.

This taxonomy has been developed progressively during periods of research carried out over the last few years. Successive formulations of this approach may be found in several of our publications.[2] An analysis similar to this was made simultaneously by the OECD and later at the UNIDO.[3]

We will now present it in a more systematic fashion, but one which essentially remains faithful to previous presentations. For each of these generic strategies, we shall indicate the goal, the means used to reach it and the underlying reasons for it.

Market Strategy

A first strategy consists in using technology licensing out not as an end in itself, but as a means to sell more products. This practice has been mentioned in many publications,[4] where it has however often been presented as a fringe benefit of technology transfer. We, on the contrary, believe that this motivation constitutes a fundamental reality – one which, far from being marginal, concerns a large number of companies, and we shall produce evidence to support this theory.

This practice is sometimes assimilated to the policy of 'loss leaders' which is used by retailing firms. As we shall see in the third part of this study, such a comparison is not altogether satisfactory. But at this point in our reasoning the reader can bear in mind that in market strategy, technology represents a kind of lever to open up new outlets. The products that the licensor thus tries to sell on the market can be of all sorts: semi-finished goods, capital goods, or finished goods, as we shall show in the following pages.

1. Product Sales The first type of sales likely to be brought about by a licensing agreement (we shall speak of induced sales), concerns semi-finished products (parts, components, etc.). So in the first stages of a transfer, whilst the licensee firm is progressively mastering the technology, it frequently does only the assembling of parts (in CKD kits) sent out by the licensor. The German firm, Grundig, for instance, undertook in 1984 to deliver 70 000 television sets in kit form in a cooperation deal made with the Chinese firm Foshan CNL/I.[5] Likewise Mafor, a small French firm specialising in locks, managed to sell 800 000 francs' worth of tools and 250 000 francs' worth of parts thanks to a know-how licensing contract with the Cameroons.[6]

These sales of intermediate products tend to continue longer than the first few years of the contract. The integration programme often falls behind schedule and the licensee is then led to continue ordering supplies from the licensor. Such was the case of Bernard, a French compressor manufacturer, whose Peruvian licensee kept up his purchases of parts for a longer period than originally planned.[7]

Moreover, such exports of parts may be reinforced by *partial licensing* of a product's manufacturing processes. Some licensors refuse to license out the most important or 'strategic' technologies, thus depriving the licensee of complete processes and forcing him to import the missing parts. Such arrangements occur as a result of the licensor's 'bargaining power', but they may equally well result from the licensee's inability to master these crucial technologies. A case in point is the technology transfer from the Berthoud company to Brazil.[8]

In selling semi-finished products, the industrial firm necessarily gives up part of its activity to the licensee, yet by means of this method, it succeeds in 'opening up' a market where the quantities sold are far greater than one might hope for in the sale of finished goods.

The licensor is in a position to capitalise on these induced movements on account of the imperfection of the market for semi-finished products. The parts and the manufacturing processes are so interdependent that the licensee generally finds it *impossible* to get supplies from any other firm than its licensor. 'There is no international market for Volkswagen doors', Vaitsos remarks,[9] except, possibly, if you are willing to buy your supplies at an extra cost – which can rapidly become prohibitive, if one thinks of economies of scale.

A second type of induced sales transfer may take place where the licensor sells the necessary capital goods. But the technology transfer may also lead to the sale of finished products when the licensee wants to broaden his range by importing those products he does not yet make. In technology transfers to Third World countries one quite often comes across agreements licensing out the less sophisticated lower range goods to the purchaser and leaving the high quality range to the seller. It was in those conditions that SNCM Brondel negotiated, in 1981, the licensing out of technologies for cable making machines for electricity distribution and telecommunications. Their partner was even considering acquiring the local sales rights for the part of the range he would not himself be producing.[10]

In all the situations we have mentioned, the sale of technologies is, in fact, a *concession* to the firm's commercial requirements. The technology is licensed out, less in an effort to 'earn money' than as a means to bolster the licensor's normal activity which is essentially to sell his products.

2. Other Forms of Market Strategy   This preoccupation also seems present in some licensing out agreements where there are no induced sales but where the licensor's objective is to *gain allies* in order to impose his technical solution on the market. For example, when Telefunken licensed out its PAL colour television patent widely in Europe, its motivation was above all to reinforce its position in the competition with the rival SECAM system.[11] This type of behaviour can be assimilated to marketing strategy, in that the licensor is not so much concerned with 'promoting' his technology as with safeguarding his industrial and commercial activity by boosting demand.[12]

What these different policies have in common, then, is the use of technology licensing as a lever for the marketing of products. Market strategies are pursued for many reasons which stem as much from environmental tendencies as from the firms' own dynamics.

3. The Underlying Reasons for Market Strategies   The existence of trade barriers between national markets constitutes an essential cause for using market strategies, since these are often designed to circumvent these obstacles. Whether those trade barriers are put up by governments or by local market mechanisms does not alter the question: with the licensing out of technology the licensor can give goods an appearance of being *'local' products*, likely to please customers and also the local government. Moreover, certain states have a policy which indirectly encourages these strategies. This is the case in most countries where industrial products are subject to lower customs duties and more lenient import quotas than finished goods; there is thus an advantage in using semi-finished goods to gain access to the market. What is more, these induced sales are known to the governments who tolerate them to the point where tied-in clauses – which are more or less illegal in almost all industrially developed countries – are apparently increasingly accepted in Third World countries.

As for the organisational basis of market strategy, it should first be noticed that the licensor finds no difficulty in making his partner accept the purchase of products when environmental conditions are favourable. Either the licensee sees that it is in his interest because his final compensation is adequate, or else he is obliged to accept because he has not mastered all the necessary techniques. In fact, it may be noticed that sales take place even when the contract does not provide for it.[13] Then the licensor's motivation is reinforced by the profits from internalisation.[14] In the case of market strategy, he may keep his core technology, go on exploiting it within the company while 'disinternalising' only peripheral technology.

Since Vaitsos's remarkable research, we know the extent of sales linked with technology transfers at an internal level: approximately two-thirds of the contracts made with Bolivia, Ecuador or Peru included a tied-in clause![15] We are probably less aware that these transfers are also very frequent in technology transfers between independent partners. In fact Lovell states that in 1969, out of 191 US licensors, 66 per cent were making profit on the sales of components and raw materials to their licensees.[16] Similarly, a study carried out by the consulting firm Conseil et Développement on a smaller scale shows that 5 technology licensors out of 8 had a continuing commercial relationship with their partners.[17]

Production Strategy

This second strategy is based on willingness to 'transmit' a technology to a firm in order to *improve the manufacturing conditions* of the product. Improvements can be made essentially in two fields: cost and quality. This means that what is expected of the technology purchaser is basically that he should take the firm's place as a processor and that he should then provide the licensor with the supplies thanks to the technology bought from him.

Again, the basic aim of the licensing is not to make profits from the technology, but to reduce costs and/or improve quality. A similar motivation can be seen in certain subcontracting relationships where the main contractor is not only a 'buyer' but also an 'instructor' who provides his partner's training before entrusting him with production operations. This transmission is quite justified in that the subcontractor has access to 'better' (i.e., cheaper and/or better quality) production factors than the main contractor. Such is the case with production strategy, the licensee disposes of *resources inaccessible to the licensor*, and that is why the latter undertakes to sell his technologies.

We shall now study the different ways in which this strategy can be used.

1. Different Forms of Production Strategy   The first type – probably the best known, and no doubt the most widespread – is that of purchasing products, whether in their finished state or not, at reduced prices. A fairly typical example is provided by Riam,[18] a French manufacturer of miniature electric motors which, since it was no longer competitive for certain products, was trying to transfer production to a licensee in a low-wage country.

The second type, probably less frequent, is purchasing from a licensee products of a *superior quality*. Indeed, it does sometimes happen that the licensee has better resources than the licensor, but he has not the necessary techniques to exploit them. This occurs occasionally in Third World countries where the exploitation of natural resources is reserved for national companies only, which may lead firms with the technology for first processing to licence it out in exchange for access to these top quality products. Philippe Koerber reports a very revealing example of this practice in the parachemical industry.[19]

What is more, over and above these questions of cost and quality, it is the *supply constraint* which governs the exchange of technology in certain industries. In the oil industry, relations between multinational companies and states have developed according to this logic; as governments have progressively taken over control of their oil resources by way of national companies, new types of contractual relationships have developed with western companies which no longer hold leases, but still have the necessary technology.[20]

Another rather unexpected form of production strategy is to *increase the product range*. This situation occurs when the owner of a technology wishes to offer a wider product range than his production facilities can cope with. He is then faced with an alternative: either to expand his production facilities or to transfer the technology to a firm which has adequate facilities. An example of this strategy is provided by the licensing agreement between LISP Machine and Texas Instruments.[21]

LISP Machine is an American firm making special computers for artificial intelligence software. It makes top quality models in limited series but has given up the idea of producing the complete range. However, the opening up of the market has made it necessary to offer less expensive models, which is where the idea originated of an alliance with Texas Instruments whereby Texas Instruments makes the machines at the lower end of the product line which LISP also markets.

To come back to what constitutes the essence of production strategy, it may be said that the licensing out of technology is only a means used to make production more competitive. That does not mean to say that this strategy is followed every time we find sales from licensee to licensor. We know, for example, that some countries (the Eastern bloc in particular) oblige their suppliers of technology to accept payment 'in kind', with so-called 'buy-back' agreements. For most licensors the 'buy-back' clause is a handicap. Yet to some it may be an opportunity to develop a production strategy. This is illustrated by the case of the Renault company in Romania, whence it imports gearboxes.[22]

2. Reasons for Production Strategies If the licensor leaves it to his partner to produce goods, and therefore to use the necessary factors of production, it is because he either does not want to – or cannot – acquire these resources himself.

It is possible that he does not wish to devote his financial resources to the purchase of material resources because, for example, he has other investment priorities. This seems to have been the case of LISP Machine in its agreement with Texas Instruments. This behaviour may also be explained by the impossibility of financing investments on account of *capital rationing*. The reasons are linked to the characteristics of the licensor and as such, are more or less independent of the environment.

On the other hand, there does exist a category of reasons which depend on the environment. We are thinking of the regulations which prevent technology owners from exploiting the resources they need. Indeed, the governments of certain countries grant access to resources on their territory only to national corporations. If we take into account the very unequal endowment of factors at an international level, firms are obliged to hand over their technology in order to have *indirect access* to the resources they need. Since the industrial policy in Third World countries increasingly requires that natural resources be exploited by local companies, there is growing pressure to license out technology in many industries. Finally the transfer of technology may constitute a means of developing *subcontracting on an international scale*. Even if the exchanges between the two partners can be assimilated to a supplier–customer relationship, there is often technical dependence which may force the licensee into a position where he is more or less the licensor's subcontractor.

This second generic strategy seems not so frequent as the former, since it is mentioned in only a few works.[23] We shall see later that it is indeed not so frequently quoted as market strategy, but it does nonetheless occupy a considerable place and therefore deserves special attention.

Technology Strategy (or Classic Strategy)

Lastly we want to consider the licensing strategy which is in fact the simplest and best known. The licensor's main motivation is simply to make the most of his technological knowledge and to get paid for it – by passing it on to a firm which wants to put it to use.

This is certainly the *main motivation* associated with technology transfer. What is more, it is quite often implicitly the only one we remember but, as we shall see later, it is actually found only in *certain situations*.

1. *The Cases of Technology Strategy*　The first – and, it must be admitted, quite frequent situation – is when the licensee is a purchaser only of the right or the information, his relationship with the licensor being merely *technical*, and therefore having nothing to do with product sales or financial operations. This supposes that the licensor has no other ambition with regard to his partner – either because he has nothing to offer, or because the licensee already has all the necessary resources. To say that this is an exceptional situation would obviously be wrong, but to claim that all licensing agreements are of this nature

would be even more mistaken. The fact is that it concerns certain technologies for certain firms.

Let us pass over the case of individual inventors for whom this is the only possible strategy, since they have no production or marketing activity. Then there is the case of so-called 'peripheral' technologies which do not fit in to the firm's *own business*. First one thinks of the technologies which are a spin-off from Research and Development programmes, and which do not always have anything to do with the firm's business sphere of activity.[24] Indeed, Research and Development functions in such a way that it does not necessarily respect the market limits, and it does happen that sometimes the results will be useable only outside the firm's business. This has led several large firms to invest in a systematic policy of economic valorisation of 'dormant technologies'.[25]

Moreover, even if a certain technique has been developed specifically for the firm's own activities, it quite frequently happens that it has *applications in other markets*. Much has been said, for example, of the spin-offs from research on ceramics done for the European space shuttle Hermes, which have benefited the automobile industry. This second phenomenon – which is specific to the procedures of Research and Development – has also encouraged large firms to reinforce their policy in this respect.

Lastly there is the case of technologies used in a firm's manufacturing process but which play only a minor part so that the transfer would not in the least endanger the firm's safety. With these *technologies of secondary importance* we find licensing agreements between competitors.

The licensing out of 'peripheral' techniques is not systematically carried out with 'technology strategy', but the latter case is quite frequent for two reasons: on the one hand, because the licensor does not run much risk, and therefore will not bother to check his licensee; on the other, because the licensor has nothing to sell to his partner, who is either in another industry, or in competition with him.

In our opinion, it does not necessarily follow that 'core technology' is never the object of licensing by technology strategy. Obviously it does sometimes happen that such strategies are put into practice. Generally this occurs when the licensee is at the same technological level as the licensor, which is the case in most exchanges between highly industrialised countries. But it is quite clear that the more 'core' a technology is, the more the other generic strategies are likely to be followed, on account of the close links with the firm's business.

Be that as it may, two other situations may be mentioned that are characteristic of technology strategy: cross-licensing and agreement following a legal dispute. In both cases, the decisive point of the agreement is 'to get the best deal out of the technology' one has mastered. Technology strategy is certainly not the only licensing strategy, but it remains a dynamic element for several reasons.

2.   Reasons for the Importance of Technology Strategy   First, the pace of *technical change* is *increasing*, so firms have to face an expensive 'technological turnover'. Secondly, *Research and Development programmes* are becoming increasingly *complex*, and require increasingly extensive resources.

So the need to pay off a larger investment in a shorter time forces one to leave no stone unturned with regard to the possibilities of valorisation, especially those which do not present any risk for the firm – e.g., 'peripheral' technology.

The development of a kind of technology 'market' contributes further to the development of technology strategies. Since the mid-1970s, increasing numbers of technology fairs serve as meeting places for those who are seeking new technologies and those who are offering them. The profession of consultant in technology transfer has been widely developed, and some of these consultancies even act as technology 'brokers'. At the same time, many data banks on available technologies are responsible for the exchange of propositions to co-operate throughout the world. Industrial firms are therefore increasingly encouraged to *transfer their 'peripheral' technology*.

After this presentation of the three generic licensing strategies, it is perhaps easier to understand what differentiates them – their objective, and the means used to attain this objective. It is now therefore necessary to evaluate the importance of each of these strategies and to identify the internal and external factors which distinguish them.

The Importance of Generic Strategies

Until now, we have had only a small number of odd pieces of information on licensors' motivations. Their interest lay in the fact that they emphasised that market strategy was far from being a fringe activity, since in almost half of the firms questioned, technology transfers were synonymous with the opening up of commercial opportunities.[26] But these were small or medium-sized firms and relatively few of them were in possession of advanced technology. So we had to widen the field and evaluate the weight of the different strategies in a more representative population. This is what we did during our enquiries into technology pricing policy.

We shall present below the results concerning (a) the frequency of each of the three strategies, and (b) their implications in terms of transactions between the partners.

The Relative Frequency of the Three Licensing Strategies

We have already emphasised that the licensor does not have one single strategy but that his motivations vary from one contract to another, and this is why we have endeavoured to distinguish the strategy followed for each contract which we have analysed. We therefore gave each subject interviewed a list of various elementary motivations behind licensing, and asked them to choose three in decreasing order of importance,[27] each of these motivations referring back to one of the three generic strategies.

From the statements made by the licensors, it was possible to assign each contract to a generic strategy. The breakdown is shown in Table 5.1.

This classification certainly shows technology strategy at the head of the list, but market strategy is also well placed. On the other hand, only one firm spoke of production strategy. Does that mean that the latter is of only marginal importance? A much larger sample would be necessary to be sure of that. In fact, results from the licensee sample would lead one to imagine the opposite. Indeed we asked the licensees to tell us what, in their opinion, the motivations of their partners were,[28] choosing from a simplified list of three motivations. The breakdown is shown in Table 5.2. It may be observed in the second sample that

Table 5.1　Frequency of the three licensing strategies (on the basis of the first motivation)

Strategy	No. of contracts
Market strategy	12
Production strategy	1
Technology strategy	20
Total	33

Table 5.2   Frequency of the three strategies according to
the licensees

Strategy	No. of contracts
Market strategy	11
Production strategy	6
Technology strategy	12
Total	29

production strategy quoted more often; however it does remain less widely used than the other two.

Be that as it may, we are not in the least preoccupied with discovering the relative importance of these three strategies in reality. At this point, it is enough for us to have found out that not all strategies are necessarily 'technological', and that the licensors therefore are not all concerned solely with making as much money as possible, with all the consequences that this implies for the technology pricing policy. Nonetheless, we have limited ourselves for the moment to the firms' statements in order to distinguish the strategies followed. This method always has the disadvantage of depending on what is said and not on verifiable facts, which is why we must validate these results by data on the transactions between partners which characterise certain strategies.

The Implications in Terms of Transactions

Market strategy and production strategy lead to product sales between the partners. The objective of the former being to open sales opportunities for the licensor, it should be possible to see a movement of sales to the licensee. In the same way, we ought to be able to observe purchases made by the licensor from the licensee when production strategy is followed.

Table 5.3 indicates, for each category of contracts corresponding to one of the three generic strategies, the proportion of them which led to sales transactions between the partners. In fact it may be noted that sales and purchases are not so frequent in the case of technology strategy as in the cases of production or market strategy. On the other hand, production strategy did not lead to purchases as one might have expected. (It is true that only one contract is concerned.)

The information taken from the licensee sample amply confirms the

Table 5.3 Frequency of the different types of transactions according to the strategies followed (licensor sample)

Types of transactions	Market strategy (%)	Production strategy (%)	Technology strategy (%)	Total (%)
Sales of capital goods	75		45	54.5
Sales of components and parts	100	100	30	57.6
Sales to make up product range	66.6		24.2	24.2
Purchases of finished products	16		5	9.1
Purchases of intermediate products			5	3.0
Others	8.3			3.0
No. of contracts concerned	12	1	20	33

first impression of consistency between the motivations and their concrete manifestation, as Table 5.4 shows. Even if each strategy leads to a certain number of transactions, market strategy is expressed by more frequent purchases and production strategy by more frequent sales. These figures contradict Contractor's statement that commercial exchanges between partners are very infrequent.[29]

The differences between the three strategies are even more apparent if the comparison is based on sales turnover instead of types of transactions, as Table 5.5 shows.

These commercial transactions naturally produce profit margins. This, then, is the real objective of licensing agreements made in the context of market or production strategy. More than the price which will be obtained from the transfer of technology, what motivates the licensor are the profits raised on the partner's sales and/or purchases. This motivation will therefore necessarily influence the pricing of transferred technology, as we shall show in Chapter 8, where we shall try to validate what has been till now only a logical assumption.

For the moment, all we have left to do is to try to reposition the licensing out of technology in corporate strategy, in order to show that this operation does not merely constitute simple opportunity seizing but that it contributes to the firm's overall development.

Table 5.4　Frequency of purchase and sales transactions according to strategies followed by the licensor (licensee sample)

	Licensor's strategy					
	Market (11 contracts)		Production (6 contracts)		Technology (11 contracts)	
	No. of contracts	(%)	No. of contracts	(%)	No. of contracts	(%)
Regular purchases from the licensor	11	100	5	83.3	4	36.4
Regular sales to the licensor	3	27.3	4	66.7	0	0

5.2　THE ROLE OF LICENSING OUT IN CORPORATE STRATEGY

The contribution of licensing to corporate strategy does not show up very well if one considers the transfer of technology from a traditional point of view. The licensor is presented as the owner of an asset which he exploits to his best advantage. The synergies that may stem from this are not obvious, nor is it possible to explain how the choice of transfers operates between the different technologies which the firm owns.

On the other hand, the approach in terms of generic strategies enables us to demonstrate the *consistency* between the firm's licensing policy and its corporate policy. The sale of technology is no longer

Table 5.5　Importance of purchase and sales transactions according to strategies followed by the licensor (licensee sample)

	Licensor's strategy		
Average value over the duration of contract (1000 fr)	Market (11 contracts)	Production (6 contracts)	Technology (12 contracts)
Purchases from the licensor	42 272.00	12 500.00	4640
Sales to the licensor	735.45	21 945.00	0

viewed as an opportunistic act, but as an element of a more *overall strategy*.

In the following pages we are going to show that thanks to our approach to licensing strategy we shall be able to locate the transfer of technology with respect to three notions fundamental to the firm's general policy: business, objectives, and international development. We shall conclude by showing the status of technology in this approach.

Licensing Out and the Firm's Business

'We are in the business of *manufacturing* and *selling* our products in many markets; we are basically *not* interested in selling our technology' (Telesios emphasis). Telesio frequently heard this sort of remark when he was interviewing company managers.[30] In our opinion this quotation is a realistic reflection of most managers' attitude, if one does not limit oneself to meeting only specialists in technology agreements, for it is more a question of general policy.

We must also clarify the idea of 'business'. On this point, we gladly join Detrie and Ramanantsoa when they define business as 'the set of know-how one must have, in the absolute, to be competitive'.[31] Performing the profession of technology selling would therefore necessitate turning it into an 'activity' and accumulating by experience the competence required. This is not within the scope of most firms because the 'know-how' specific to this type of operation is quite *distinct from the firm's usual activities*. Information must be negotiated – instead of negotiating products, knowledge must be passed on, and not only goods, plants must be launched, and not only managed ... Of course these skills are not inaccessible, but in specialisation choices firms prefer to put to better use what they know already rather than launch themselves into a new business. The more so in that, as the internalisation theory teaches, the information 'market' is such that what can be taken out is generally less than the supplier could himself take out. This is why the transfer of technology most often appears as a fringe activity in comparison with a firm's basic business. This can easily be verified by observing how small a part is played most of the time by technology payment in the firm's turnover – at the most, a tiny percentage.[32]

It is therefore quite understandable that firms should frequently be reluctant to part with their core technology insofar as this is what constitutes the mainstay of their profession. They feel better able to exploit it, and are afraid of its being inefficiently used if they pass it on. Of course, they can always resort to outside help, and avail themselves of specialised consultants to assist them in this task. However, they will have to take on a major role in *servicing the transfer* as regards formalisation of the technology, possible adaptions, training of personnel, technical assistance, etc.

Technology Transfer May Also Reinforce the Business

Yet in spite of these obstacles, some owners of technology do engage in an active licensing policy. Unless one imagines that these companies are behaving illogically, the possible link with their business should be examined, and that is made feasible in terms of generic strategies. Indeed it may be noted that, if we drop the traditional notion of technology transfer as a mere handing over of the firm's knowledge, licensing may appear as a means of reinforcing the licensor's business. With a market strategy and the implied sale of products, the licensor continually *exploits his skills*: he agrees to give up part of his activity but concentrates on what makes his business special, whilst keeping mastery of his core technology.

With a production strategy and the purchases it leads to, the firm's business is not neglected either, since the licensor considers himself in a weak position for part of the manufacturing process which he is going to entrust to his licensee. It is because part of the technology is beyond the scope of his business that he transfers it. But he *retains a place on his market*, and thus a commercial know-how which he will continue to use thanks to a licensing agreement.

Finally, certain forms of technology strategy may also reinforce the licensor's business – like cross-licensing, for example, where the main motivation of technology licensing is the *acquisition of another technology* which will be added to the firm's know-how. But there is a whole series of licensing agreements which are beyond the scope of the firm's business, all of which fall within the province of technology strategy; these are 'peripheral' technologies, by-products technology or agreements after a prosecution for technology emulation. Very often in these cases the 'synergy' with the business is almost non-existent.

We thus find a wide range of synergy between the policy of technology licensing and the licensor's business. This remark should encourage technology supplier firms and recipient firms to consider licensing within the perspective of general policy and not as an end in itself. By seeking to reinforce his business, the licensor could better achieve his aims.

Licensing Out and the Firm's Objectives

What we understand by 'objectives' here is the type of long-term performance desired. Corporate objectives have been the subject of long theoretical debates which we do not intend to enter into here.[33] Our ambition here is limited to pointing out that technology transfer may contribute to attaining the different objectives defined by the theoreticians.

The generic strategy approach enables us to do that. However it would be a long business to treat all the theories of the firm's objective, so we shall restrict ourselves to the best known. The contribution is pretty obvious, if we define the objective as the *maximisation of profit*. The three generic strategies can play their part here. Unfortunately this concept of the firm's objectives suffers from theoretical and practical insufficiency.

If we consider another approach, *maximisation of sales* or *growth*, which has experienced a certain success with the work of Baumol and Penrose amongst others, the link with licensing strategy is more tenuous. Whilst retaining a classical concept of licensing, one cannot in fact explain this link satisfactorily. Technology licensing out appears as a 'marginal' operation, in that it does not directly contribute to the firm's progression towards its objective. On the other hand, with an analysis in terms of generic strategies, a possible consistency with the objective may be seen. It is perfectly clear for market strategy, and almost as obvious for production strategy, when the licensor needs the licensee's resources to keep or develop his market, thanks to cheaper products.

A final concept of the firm's objectives may be considered: the *reduction of uncertainty*. This theory has been developed by Galbraith, Caves, and Jacquemin,[34] and states that the objective of the firm's leaders is to reduce their risk and avoid the hazards due to the evolution of the environment, by increasing their capacity for controlling that environment. The classic view of licensing makes it difficult to understand how the licensing of technology should contribute to increasing power. If the licensee is merely a purchaser, the relationship with him will be nothing more than contractual and the only pressure on him will not have more than a strictly legal dimension. Generic strategies, however, show us how licensing helps reduce uncertainty. As we have seen, market strategy is a means of making outlets 'safer', since they would otherwise tend to disappear or decrease because of government decisions or the activity of competitors. Similarly, production strategy aims to make the licensor's supplies 'safer'. Needless to say, licensing strategy does not ensure total safety – no more of course than other actions of the firm on its environment, since, as A. Jacquemin has it: 'A firm's power is always threatened and always open to reappraisal'.[35]

There are many other objectives in a firm which we cannot deal with, but we consider that in most cases the *synergy offered by licensing cannot be explained by the classical concept* of technology transfer. To conclude, we prefer to draw attention to a particular sub-objective created by international development.

Licensing Out and Internationalisation Strategy

Technology licensing is not in itself an operation of internationalisation, since this type of transaction also takes place within national frontiers, but such national transfers are more infrequent if one considers the licensing arrangements made by manufacturing firms, on account of the risk of increased competition. For these firms, licénsing appears mainly as a form of international development.

A large number of publications have considered the relative advantages of the three forms of internationalisation: exporting, direct foreign investment and licensing. Now we have built up a more accurate analysis of licensing with generic strategies, we have at our disposal a new framework of comparison. Indeed, it may be noticed that a *relationship* exists between the three forms and the three strategies.

Thus exporting is linked with market strategy, since its aim is essentially the sale of products. Foreign investment corresponds either to market strategy or production strategy. Market strategy answers to the objective of gaining new markets through relay subsidiaries whose job is to produce and sell locally, and which Charles A. Michalet has so well described in *Le Capitalisme Mondial*. Production strategy rests on an objective of obtaining supplies at costs, and in quality and quantities, required by means of workshop subsidiaries. Here we find Michalet's production and supply strategies. He attributes another strategic alternative to the multinational firm, technology strategy, which in fact is not concerned with direct foreign investment, but with technology licensing.[36] In our opinion, this form of internationalisation permits only the first two generic strategies. On the other hand, technology transfer *adds this dimension whilst retaining*, as has been shown, *the first two*.

Finally, a certain progression may be observed among the three forms of international development, as illustrated by Table 5.6.

To conclude, we would like to emphasise the status attributed to technology in licensing strategy.

Basically it does indeed constitute an element of the firm's market power, as most authors agree. However, exploitation of this power may take two different forms:

- Either by *direct deduction* of part of the rent that the licensee gets from his market. This is the case of technology strategy or *'pure licensing'*, when the royalties and other contractual returns alone allow this deduction.
- Or indirectly, by the *lever effect* which ownership of technology confers on commercial transactions with the licensee. In this case, which corresponds to market and production strategies, the technology returns not only function as a deduction from income, but they should favour the development of sales and purchasing operations between the partners. Technology is no longer 'dis-internalised' to use to Dunning's expression.[37]

On the contrary, such technology returns allow transactions with the licensee for semi-finished or finished products to be *partially internalised*. It is therefore possible to speak of *'quasi-internalisation'* as a designation of this second form of exploitation of the market power conferred by technology. One may even wonder if it is not basically a *fourth form of international development*, intermediary between direct foreign investment and pure licensing.

We shall revert to this point in detail when we study the relationship between licensor and licensee (Chapter 7), and again when we reach the

Table 5.6 Ways of internationalisation and generic strategies

	Market strategy	Production strategy	Technology strategy
Export	×		
Direct foreign investment	×	×	
Licensing	×	×	×

recommendations to firms who commit themselves to a procedure of internationalisation (Chapter 9). Our models of pricing policies will naturally have to integrate this new concept. Before that, we will try to show that the licensee is not a mere 'buyer of technology', by analysing the strategies of acquisition.

6 Licensing In Strategies

An agreement to licence out technology is the result of two parties wanting to do so, and it can be effective only if both partners think it will be profitable for them.

It is therefore useless to analyse the strategy of the licensor without also studying the behavioural logic of the licensee. Indeed, in order to demonstrate the supplier firm's motivations, we must show that his expectations may be satisfied by the acceptance of the recipient firm: without this, we shall not be able to understand the strategic dimension of licensing. It is true that the licensee may seek only to acquire technical skill and refuse to make any other transaction with his 'supplier', whom he then confines to 'technology' strategy. However, quite frequently this is not the case, because the licensee's behaviour is *not* merely that of a *purchaser of technology*.

To understand this, we must first examine the licensee's own motivations (Section 6.1), and it will be shown that this is not the only factor that determines the type of agreement he wishes to conclude. The resources available to the licensee must also be taken into account – for, to a considerable extent, they determine his freedom of manoeuvre when faced with his partner's proposals (Section 6.2).

One of the fundamental ideas in this chapter will be to expose the commonly-held misconception which claims that the licensee is systematically dominated in technology transfer agreements. We shall show that this depends on the state of the resources at his disposal and that various different situations can (and do) occur.

6.1 THE LICENSEE'S MOTIVATION

Less research has doubtless been done on motivations for acquiring technology than on the ones for licensing out, possibly because the point seems 'obvious'. Unfortunately, this is not so; it is sufficient to refer to the few works devoted to the subject to be convinced of this. A large number of reasons for acquiring a licence may be found, for example, in Lovell's work.[1]

On reflection, and after a good deal of discussion with licensees and specialists,[2] we consider that *two levels of motivation* should be distinguished. The first level concerns underlying motivation – i.e., the market – but it must be made clear why technology acquisition has been chosen as a means of *access* to the market. We shall then speak of specific motivations. Finally we shall consider the motivations a licensee may have to obtain *complementary resources* from his partner, thus going beyond his role of mere purchaser of technology.

The Basic Motivation: Market Access

Directly or indirectly, the decision to acquire technology always stems from market considerations; no one would acquire technology without intending to use it. On the other hand, it is well known that technology may be produced and then not used. We shall not here consider the case of dominant firms who buy up patents that may one day prove a nuisance in order to 'sterilise' them. Apart from this exception – the actual importance of which has never really been measured – companies acquire licences to *build up their presence on the market*, even if at first they serve only to 'unblock' the process of research and development for the new product or process. It is interesting, however, to distinguish between the case where the licensee is already present on the market and that where he is seeking to diversify. The latter generally seems to incur more frequent acquisitions of technology.

Building up Presence on Existing Markets

The acquisition of technology to reinforce the position of the company on a market where it is already established is often depicted by the case of the 'follower' who is trying to catch up with the 'leader' of the industry. A good illustration of this point is to be found in European electronics firms with their agreements relating to video cassette recorders. Business strategists have devoted a lot of attention to this case, with the related problem of technology policy choice.[3]

1. The Licensee is not Always a Follower   It should be recognised that the licensor is not necessarily the market leader, but may on the contrary be a fringe competitor who has had the good luck to perfect highly useful technology which he cannot impose on the market himself because of his weak commercial position. He may therefore decide to entrust his innovation to one of his main competitors who may be better able to exploit it.

But William Rothschild also indicates (and this is fundamental) that the origin of the innovation may be external to the industry.[4] Improving one's competitive advantage by acquiring a licence does not necessarily depend on the good will of the pioneer; fortunately other supplier firms exist, such as research centres (Battelle, for example), individual inventors and firms belonging to other sectors of activity which have developed technology applicable outside their own sphere of activity.

Finally, the licensor may be working in other geographic markets, whether he be a leader or not. Under such circumstances, a good position on his original market would seem to be an asset, because out of 33 contracts in our sample of licensors, 63.6 per cent were leaders, 24.2 per cent were challengers and 12.1 per cent had a fringe position on the market.

2. Reasons for Licensing In Whatever the status of the supplier firm, however, taking on a licence is the result of more or less long-term commercial motivation which, naturally enough, may take different forms:

– *Improving the quality* of existing products, for example by modifying certain components. This may be illustrated by the introduction of the anti-blocking braking system on the Renault 25.
– *Launching new products* on a market which is already occupied. The licence then no longer covers components, but the complete product.
– *Increasing efficiency* in cost and/or quality of the manufacturing process. To illustrate this, the licences granted by Pilkington Glass on its 'float glass' process have given a number of producers the means to make considerable progress.[5]
– *Penetrating new geographic markets* which call for the firm's adjusting to technical standards or even, quite simply, obtaining the patentee's permission to market the product. The sailboard industry provides a particularly interesting example of this point. The patent was originally taken out in the United States by a Mr Hoyle-Schweitzer, who applied for protection in almost every industrialised country except France. However it was France which subsequently became the largest market in the world and the world market leaders originated here: Bic, Tiga and to some extent also the Swiss firm, Mistral. However, when these firms wanted to sell in markets protected by the patent, they were obliged to pay royalties to Hoyle-Schweitzer, which they did not do in France.

- *Taking advantage of the market protection* that the governments apply in favour of local manufacturers, particularly in developing countries. The growth of companies, by having recourse to the technology of foreign 'competitors' is facilitated by the fact that entry barriers prevent competition on the market.

In all these situations, the justification for taking out a licence is to obtain access to a market – generally by means of improved competitiveness, but also by paying an entry fee. Nevertheless these strategic moves take place on the market, or the product/market, over which the firm already has a certain hold. It is quite different in the case of diversification.

Diversification to Penetrate New Markets

It would seem that the desire to diversify is an important factor in the acquisition of technology.

In this context, we could take another look at the edifying analysis of this phenomenon by Caves, Crookell and Killing[6] observing the frequency of licences granted in a sample according to the degree of diversification. They here distinguish four levels of diversification:

- *Existing product* (therefore no diversification): the licensee already masters all the skills, including product design, manufacturing processes, and marketing.
- *Closely related diversification*: the firm already masters skills similar to those required for the new product.
- *Loosely related diversification*: the licensee is lacking in one or two of the skills identified above.
- *Unrelated diversification*: the licensee does not master any of the skills necessary for launching the product on the market and will have to learn them by acquiring a licence.

1. Licensing In as a Means of Diversification By asking the 80 licensees in their sample which of the three skills necessary to exploit a licence (product design, manufacturing processes or marketing techniques) they had already mastered prior to the agreement, Caves and his colleagues were able to situate the contracts in one of the four levels of diversification. They found that almost 80 per cent of licensing agreements corresponded to a choice of diversification. Our own observation fully confirms this sort of figure.

The *licence/degree of diversification relationship* is nevertheless not simple as the detailed breakdown (Table 6.1) shows.[7]

Indeed, it seems that if the development of new activities generally leads to the taking out of a licence, the distribution between the three types of diversification is very uneven. This is most probably due to the fact that diversification is all the less frequent where it concerns activities that are remote in terms of necessary skills.

Licensing in may thus be analysed more as a search for complementary skills than as the acquisition of a complete skill. It follows that the potential licensee has useable capacities available, together with the acquired technology. In other words, the justification for licensing in often rests on the possession of a competitive advantage (skills relative to the product, to the process or to the market) that the firm acquiring the technology is seeking to exploit better by adding the complementary skills. The licence acquired by Texas Instruments from LISP Machine is typical of this approach. But whether penetrating a new market or increasing one's presence in the existing market, it is important to grasp the influence governments may exercise as well.

2. *The Influence of Governments Over Technology Acquisition* The transfer of technology may concern a number of *objectives* in the context of *industrial policy*:[8]

- To expand gross domestic product.
- To improve the balance of payments.
- To increase employment.
- To raise revenue and purchasing power.
- To increase national independence.

One may be tempted to consider that concern for the market disappears under such circumstances. But the truth of the matter appears more complex. If we exclude technology imports for military purposes,

Table 6.1　Licence/degree of diversification relationship

Degree of diversification	No. of contracts
No diversification	18
Closely related diversification	22
Loosely related diversification	34
Unrelated diversification	6

it seems to us that the market and the needs of the 'consumer' remain the underlying motivation for these operations, even though it is the governments who impose their vision of what the market expects. Although the decision to acquire a licence may subsequently prove commercially unjustified, this does not mean that there was no market motivation originally. One may imagine, at most, that this motivation was badly treated. It remains that the acquisition of a technology, as in the other cases, is aimed at its commercial use, more or less directly and more or less efficiently.

However, recourse to a technology developed by another firm is only one of the ways of reinforcing competitiveness and diversification. Internal development remains a possible alternative. The justification for acquiring a licence can thus be complete only if we also take into consideration the specific motivations in favour of licensing in and against internal development of technology.

Specific Motivations: Economy of Technological Resources

The essential advantage of acquiring a licence over internal development consists in economising resources. The licensee, seeking to reduce the cost of developing the technology he needs, chooses this means every time its cost appears less than that of any other alternative. In fact, behind this rather general motivation, four main reasons may be found which explain the licensees' decision.

Absence of Research and Development Capacity

The first reason is the licensee's inability to do Research and Development himself. We know that some firms do not have a Research and Development department, even in the most advanced sectors of activity. This results from either the absence of human resources with the necessary qualifications and the scarcity of such resources on the job market, or an insufficiency of financial resources, or again a political choice. In the North–South transfers we often come across this type of situation where the licensee is a relatively young firm which prefers to buy technology rather than *devoting means to developing technology* which may already exist, and which has perhaps even been available for years.

The Desire to Avoid Costs, Delays and Risks of Research and Development

Even when a company possesses the necessary skills for developing technology internally, it may refuse to launch out on a Research and Development programme. Not only does this programme represent considerable expense (engineers' and technicians' salaries, cost of materials, manufacturing and prototype testing costs, etc.) but it imposes *delays and risks*. In comparison with existing technology which a licensor might be willing to grant, the time required to set up an in-house Research and Development programme is longer, and might thus lead to a substantial opportunity cost because of later entry into the market. What is more, Research and Development activities are characterised by a degree of uncertainty which means that the firm is never sure of the quality or the timing of the results. The licensee may thus prefer to buy technology that he will be able to use more rapidly, since the time factor involved in transfer is usually shorter than that for development.

This decision may be analysed as a form of *externalisation of Research and Development*. One quite often comes across licensees who make a cost–benefit calculation of licences compared to in-house Research and Development. For a number of years a French refrigerating equipment company has thus adopted the position of licensee and, in a manner of speaking, it 'subcontracts' its Research and Development out to an American firm.

The Desire to Complete the Results of Research and Development

When a firm not only has Research and Development capacity at its disposal but is actually engaged in programmes concerning the new products it wishes to launch on the market, it does sometimes also acquire licences in the same field to complete its own results. Indeed the development of an innovation is often not a matter of mastering a single technology, but frequently involves a combination of different technologies. If the firm is not able to develop them all at the same time, it has to make the choice of *acquiring those it lacks*. In fact, it may be noted that the firms which are the most advanced in innovation – like Texas Instruments and Thomson, for example – devote considerable budgets to licensing in: a first sign that shows the acquisition of technology does not necessarily mean 'being a technology latecomer'.

The Desire to Avoid Counterfeiting

A firm may be obliged to take out a licence even when it has all the necessary skills (product, process and marketing) for launching a new product. Indeed, having this knowledge and these skills does not mean one is the 'owner' in the industrial ownership sense, since other firms may have prior patent rights for the technology. The fact that a firm has entirely *rediscovered a technology* in good faith and without any external assistance makes no difference: it could be charged with counterfeiting if it exploits it without permission from the patent-holder. The arrangement between Software Arts and Visicorp for the marketing of Visicalc is a good illustration of this type of situation.[9]

The constraint of industrial property may be less direct. It is possible, for example, that a technology invented by a firm may not be identical to the one that has been patented, but that certain stages of the process may be contentious. In the field of antibiotics, Telesio shows that there are a number of cases of *intrusion by one patent on another*, since they are more or less built up on each other.[10] The exploitation of this technology would become very difficult were it not for cross-licensing on these interlocking patents.

This motivation – where the aim is avoiding a lawsuit for counterfeiting – is a good example showing that the acquisition of technology does not always concern firms that are weak in the field. It also emphasises the distinction to be made in a licensing agreement between the transmission of information and the permission to use this information. In the second case it is a question of acquiring the right to use it, since the licensee is supposed already to be in possession of the information necessary to produce and market the product. In other words, the communication of know-how is not envisaged, no more indeed than for the preceding motivation where we stressed the cost of Research and Development.

Now the content of technology transfer is not limited to the results of Research and Development, it also includes experience. This may give rise to a final motivation.

The Desire to Speed up the Learning Process and the Accumulation of Know-how

This motivation concerns the acquisition of practical information about manufacturing processes, and in particular how to carry out the different steps of the production process. One might see a similarity between this motivation and the one connected with the Research and Development time factor and risks. We prefer to make specific mention of it because it seems that this learning process is radically different from that of Research and Development. In Research and Development, a firm can set objectives, finance the operation, follow a strategic plan of action. On the other hand, know-how *takes time to accumulate* as experience grows and opens the way for individual improvements which are more or less important and more or less unexpected.

A company that wants to launch a product which is new for it but which another firm has been manufacturing for several years will come up against this problem. Either it decides to follow this learning process and to be satisfied with rather poor quality and productivity, improving only very slowly, or it is prepared to buy the know-how, the practical tips and experience and thus considerably speed up its progress along a path which will have to be followed anyway. The role of technical assistance and on-site training is decisive in this case.

So these 'technical' motivations are related to the means of satisfying the most fundamental motivations – for the licensee – access to the market.

Derived Motivations: Access to Complementary Resources

Emphasising the use of technology for commercial ends does not necessarily mean that the licensee is nothing but a purchaser of technology. Quite frequently the licensee's request exceeds the mere acquisition of technology and includes *complementary transactions* with the licensor.

Indeed, it would be a mistake to imagine that these transactions only satisfy the licensor; quite the contrary. As Yong Hee Chee shows in Table 6.2, appreciation should be much more qualified.

First of all, the buyer, too, may derive benefits from the purchase of raw material or semi-finished products from the licensor. The size of his market, for example, may prevent him from integrating the process completely, because some of the equipment would be insufficiently used. Or, again, some sorts of specialised labour may not be available on the local market and the licensee may be content to obtain the necessary components so as to avoid the problem of recruitment. These are only two examples, amongst a host of others, which all show that over and beyond the acquisition of technology the recipient firm's request may cover a wide ground. However, we should not overlook

Table 6.2   Possible benefits and costs of various inputs tied to technology

Major categories	Possible benefits to licensee	Possible costs to licensee
Raw material intermediate goods and equipments	Access to source of supply	Overpricing of tie-in goods
Equity capital	Risk sharing by initial capital contribution	Dividend, and implicit costs due to loss of control
General management	Business know-how and management techniques	Reimbursements in the form of salaries and other personal expenses; implicit cost of inappropriate management method
Finance	Availability of funds; access to international money market	Interest and other service charges
Marketing	Know-how, management skills of marketing and selling; access to foreign markets	Fees, loss of direct access to foreign and cost due to lower transfer prices

Source: Yong Hee Chee (1979) p. 17.

the risks and costs of this first type of transaction with the licensor, and in particular the overpricing of products purchased.

Next it is obvious that *possible sales to the licensor* may present some interest for the licensee. These would be complementary to the possible outlets on his local market and also provide a number of marketing skills, thus enriching the contents of technology transfer. At the same time one must be aware of the concomitant risks, such as an insufficient transfer price or the loss of control on foreign markets. The *other transactions* (equity capital, general management, finance) are also seen to be ambivalent, which helps to confirm the idea that the acquisition of a licence cannot always be classed as the pure purchase of technology.

Finally one must add to the first motivations (responding to the markets and economising on the development of technology) the search for complementary inputs which the licensor may provide. The first two types of motivation are present in all licensing agreements; these may be joined by the third category according to the licensee's priorities and resources. So these three levels of motivation are not sufficient to determine the content of a contract acceptable to the licensee; for that, it is necessary to take into account the structure of his own resources.

6.2 THE LICENSEE'S RESOURCES

The firm which takes steps to acquire a licence implicitly admits that it does not possess certain resources. These are probably resources of a technological nature, although they may be of another type. Theoretically, the greater the lack, the more resources the firm will have to obtain, but nothing forces the future licensee to apply to a single licensor to get what he needs. If other firms in his immediate environment can provide him with these resources, we may imagine that the licensee will apply to them, if only to avoid too great a dependence on his partner.

An examination of the resources available to the licensee should not therefore be limited to what he actually possesses. The resources of the *local environment* should also be counted, particularly when the government restricts imports, as is the case in most developing countries.

The first type of resource is linked to the licensee's financial capacity, as Yong Hee Chee's survey shows. But this point will not be emphasised insofar as financial resources do not impose strict complementarities in relation to the technology employed. On the other hand, we will pay more attention to technological and commercial resources, which seem to be more crucial factors. Mention will also be made of an indirect resource available to the licensee, namely that there may be competition between several potential licensors.

Thus it may be shown that the sum of resources available not only influences the content of the contract, in terms of induced services and transactions, but also decides the extent of the recipient firm's bargaining power.

Technological Resources

The general rule here is: the wider the licensee's technological skill, the less need he will have of major services (technical assistance and training). He might even be satisfied by a simple transfer of rights. He might be in a position to refuse complementary purchases, thus imposing a technology strategy on his licensor.

Now, contrary to popular opinion, the licensee is *not always* technologically backward since, as Mansfield shows, firms spending large sums on Research and Development tend to be quicker in adopting technology developed by others.[11]

Our own observations are in complete accord with this statement, since the 'biggest' sellers of technology we have met are generally also – in terms of the number of contracts – the 'biggest' buyers. It is obvious that for such firms the content of the transfer is more limited insofar as they do not have to rely on their licensor for all the necessary technology: core technology accompanied by complementary technology. Thanks to their resources, they are not obliged to assume a position of dependence on the licensor.[12]

For all that, we have to admit that in the studies we have undertaken among licensors and licensees, the latter show a tendency to devote less of their budget to research and development than the former: on average 1.67 per cent as opposed to 4.01 per cent. But it is true that the sample of licensees included a large number of Moroccan and Portuguese firms, whereas the other sample was entirely made up of French firms.

The demonstration of a relation between the content of transfer and the complexity of a technology seems to be valid only if, like Lassere and Se-Jung Yong's study, the observation is limited to a single country and thus to a particular technological environment.[13] The extent of the services and transactions between the partners is in fact determined by the relative extent of their resources, which are themselves obviously influenced by the local environment.

Commercial Resources

By 'commercial resources', we mean essentially knowledge of the market in which the technology is to be exploited. This knowledge is naturally not unrelated to the licensee's position in his market, since in this domain skill depends mainly on experience and the distribution infrastructure (sales force, relations with distributors, customer-files, etc.). Commercial skills represent a double asset for the licensee.

– *First asset*: the licensee will be in a better position to *evaluate the potential profitability* of the technology on his market since he will be able to estimate the size of the market, the possible market share, the margins to be allowed to the distributors, the final price, etc. Now this skill is a prerequisite for an effective bargaining power.[14]

Indeed, Yong Hee Chee shows that the purchase of technology may be analysed in terms of sharing out future revenue. Now since the aim of negotiation is to keep the largest possible share of the future profit, any extra information about this strengthens the negotiator's position.[15]

- *Second asset*: the licensee *may bring a skill* which his partner perhaps lacks, and in that case the latter may be prepared to make concessions. In particular, in such a situation, the licensee will not need orders from the licensor. In other words, if the purchaser of technology is already well established in his market, the licensor's 'production strategy' may well be hampered. Again, the quality of the licensee's resources may force the licensor to use technology strategy, unless he is making a deliberate strategic choice.

In fact, the licensee often does have a good position on his market since, according to the licensors questioned during the survey in France, the 33 licensees were classified according to their market position as in Table 6.3.

According to these figures, almost 58 per cent of the purchasers of technology have some sort of commercial skill, at least in the eyes of their supplier firm. On the other hand, the number of 'beginners' is impressive – indeed, it is the largest single category. In other words, licensees' *commercial skills are very unequal* – sometimes these skills are non-existent, but often they are considerable. It all depends on the degree and the nature of diversification that technology transfer implies for the licensee.

Moreover, it is quite common to find that a *relationship already exists* between the two partners: either because the licensee was already a local distributor for the licensor, or because the two firms had some other form of relationship (financial, technological or personal.) This represents about one-third of the cases among the licensees and more

Table 6.3   Licensees: Market position

Market position	No. of firms	(%)
Leader	10	30.3
Challenger	9	27.3
Marginal	3	9.1
Beginner	11	33.3

than 12 per cent of the contracts in the case of the licensors; these results are confirmed by a previous survey where 14 per cent of the licensees were former agents or local distributors for the licensor.[16]

Finally the role of governments in technology transfers is not neutral, since they can *reinforce* the bargaining power of the purchaser thanks to information at their disposal, and also by granting the licensee protection from competition. These state interventions cannot of course give commercial skills to the beginner licensee; however, they do act as a substitute since they make his task easier during the negotiations insofar as the purchaser receives all the available information and suffers from no local competition. It often happens, then, that the negotiations favour the purchaser, but it must not be forgotten that this is because of a substitute for commercial skills which will be worthless when activities really begin.

At the same time, the licensee's bargaining power is strengthened if several potential licensors are competing for his attention.

'Competitive' Resources

If there is rivalry between several candidates for the licensing out of technology, the situation is *favourable* to the licensee. There are several reasons for this.

First of all, of course, the classical competition mechanism tends to eliminate super-profits and bring the offer price down to the cost of production plus a 'normal' profit margin. This mechanism never works completely in technology transfers because that would mean that the technology concerned would be quite commonplace – that is to say, mastered by a large number of firms – and that almost all of those firms would be aware of a demand for this technology. These conditions never occur, except perhaps in operations which are closer to the provision of services (e.g., harbour developments) than real licensing out of technology.

A second reason is linked to the spread of information. We know that access to technical and commercial information is a deciding factor in bargaining power. Now, rivalry between potential licensors causes spreading of information, since each one wants to prove his superiority with regard to his competitors.

Let us mention one last reason (although there are yet others): competition has the excellent result of showing up the differences, not only between the technologies offered, but also between the complementary services proposed. In this way the licensee can better distinguish in these 'proposals' between what has to do with the nature of the technology in question and what is due to the strategy pursued by the licensor. This distinction is a prerequisite to a rational choice in the type of agreement into which the licensee will accept to enter.

However, these reasons are valid only provided that there is a certain degree of competition. Not much data was available on this point up till now, but our survey has shown that there is often some rivalry between the licensors who are candidates for the licensee's favours. Competition exists in 50 per cent of all cases according to the people we questioned. Even if this data is only approximate, given the size of our samples, it does suggest that licensing out cannot be analysed in terms of bilateral monopoly. The licensor and licensee are not in a position of total monopoly and monopsony. A certain rivalry appears between them, and this necessarily affects the bargaining power of each party.

Specificity of Competition Between Licensors

It should be noted, moreover, that 'technology competition' – the rivalry between licensors – is of rather a different type from the competition in selling other products.

Firstly, licensing out does not work with the same competitive advantages as the market of goods and services. Competition is less symmetrical in the sense that small firms in a given sector may succeed as well as – or better than – large firms. A good example of this is the case of the French compressor manufacturer, Bernard, a small company with a workforce of only of 260 representing at the time a mere 15 per cent of the French market. In the face of competition from the vast multinational corporations in this industry, Bernard carried off the invitation to tender sent out by Peru because their technology was probably well adapted, but also because the company provided a more flexible negotiating partner than the industrial giants. Many cases like this are quoted in the Weil report.[17]

Secondly, the 'sellers' have very *different* objectives. Not all the competitors on the markets for goods and services have the same objectives, either. Some are trying to maximise their profitability, others their growth rate, still others are trying to preserve their financial independence. But as far as exchanges of technology are concerned, the differences are even more marked. This is the result both of divergences in corporate objectives (those previously mentioned) and differences as regards the licensing-out strategies. Beyond the real importance of these differences in objectives, which we are at present unable to measure, is the influence which the situation may have over the type of agreement that the licensee will finally sign, and especially the price which will be set. On account of the diversity in strategies, and therefore in pricing policy, which (as we shall show later), stems from them, we can expect that the market price will *differ considerably* from the *theoretical price* which would have been found had all the agents followed the same objectives.

The Relationship Between the Licensee's Resources and the Type of Agreement

As a conclusion, we can now advance a few results which give sufficient weight to our hypothesis of the influence of the licensee's resources on the type of agreement. Table 6.4 below indicates that the more limited the technical resources are (Morocco and Portugal compared with France) the more complementary transactions accompany the licensing out. In the same way, sales to the licensor, much more evident for the Moroccan licensees than for the Portuguese and the French, are due as much, doubtless, to the limited nature of their 'commercial resources' (limited size of the Moroccan market) as to the low labour costs from which they benefit.

If our last remark is valid, this means that the payment and transaction structures therefore represent an expression of the *state of resources and motivations of the licensee*, who wishes both to compensate for his lack of resources and to make maximum use of those he has at his disposal (labour in the case of Morocco). At the same time, it sheds light on the use made of the licensors' three generic strategies, with their partners' consent. This also explains why knowledge of the licensor's strategy is essential for price negotiations. This point will be

Table 6.4   Mean technology payments and induced transactions according to the licensee's country (licensee sample, 33 contracts, 1000 francs)

Transaction	Morocco	Portugal	France
Technology payments[1]	4002.7	4768.8	21081
Sales to licensor	9512.1	823.7	0
Purchases from licensor	27994.0	16833.0	10743

Note:
1. Definition of this term is given in Chapter 2.

further developed later, when we present the models of price calculation from the licensee's point of view.

Now that we have presented an analysis of the licensor's and licensee's strategies separately, we can tackle the last stage and consider the relationship between licensor and licensee. This stage is indeed necessary before starting on the question of prices, since the policy applied by licensors implicitly includes the nature of this special relationship between the two firms.

7 A Theory of the Licensor–Licensee Relationship: the Quasi-Internalisation Model

The most suitable framework for analysing the relationship between licensor and licensee is without doubt what is commonly termed 'institutional economics'. This refers to a line of research that is concerned with the running of businesses as both institutions and as agents on the market. According to this approach, the organisation of industries (level of integration, degree of concentration, barriers to entry, etc.) is governed to a great extent by institutional factors and not by supply and demand features.

Research in this field began as far back as 1937 with R. Coase,[1] but was widely developed in the 1970s by Williamson,[2] Alchian and Demsetz[3] and Casson and Buckley;[4] the latter examined its international implications. The conceptual framework conceived by these authors is of obvious interest when observing the 'markets' for technologies. If these markets were indeed perfect (as would be the case if the seller had a total monopoly, if there were perfect competition between competitors, perfect information and no transaction costs) then the traditional theories regarding the market economy would be adequate for understanding the licence relationship. However, as was pointed out in Chapter 3, exchanges of technology *do not obey these perfect rules*. On the contrary, serious 'imperfections' exist: in particular, there are high transaction costs. Why, then, do such transactions take place?

The explanation offered by supporters of the institutional economics theory is as follows. Because of the imperfections that exist on the intangible assets market, owners resort to either of the following courses of action:

– They *refuse to sell them*, as they would not receive sufficient returns, and prefer to continue exploiting their assets to the full themselves.

This is the internalisation theory put forward, notably, by Buckley and Casson.

- They *agree to license* or externalise them, *while at the same time imposing certain restrictions* on how the information thus made available is to be used. Restrictions of this kind are not to be found in any other type of market. They concern the territory where the technology is to be used, non-exclusivity, type of application, etc. This second aspect of the theory is analysed in particular by Teece[5] and by Caves, Crookell and Killing.[6]

This 'institutionalist' alternative between internalisation and dis-internalisation underlies most of the empirical works devoted to the licence relationship, as is demonstrated by Bonin in his review.[7]

However, it is our contention that a middle road exists between these two, namely that of '*quasi-internalisation*', as it would appear that externalisation occurs only in certain cases of licensing, where the object of the licence is purely technical and no other transaction is involved. This corresponds to the technology strategy.

In contrast, we shall attempt to demonstrate that in the case of technology transfers which are accompained by commercial transactions between the partners, there is only partial 'dis-internalisation'. Continuing the line of reasoning adopted by Caves, Crookell and Killing, we shall put forward the hypothesis that imperfections in the technology 'market' lead licensors not only to impose restriction clauses but also to introduce other transactions into the licensing agreement.

Quasi-internalisation – this intermediate form between the market and internal organisation – was envisaged by the main promulgators of the institutional economics theory:

- Coase indicates that when resource management depends on the buyer, it is possible to speak of internal organisation of a firm.[8]
- Williamson speaks of 'non-firm' forms of internal organisation.[9]
- Casson and Buckley explain that multinationals sometimes limit the internalisation of their activities so as to allow each individual unit to buy and sell outside the group.[10]
- Houssiaux may also be associated with this line of research, insofar as he considers subcontracting a form of quasi-integration which modifies the behaviour and performance of the partners involved.[11]

In spite of these precedents, there is little analysis of quasi-internalisation in technology transfers. This would nevertheless appear to be an interesting line of approach that needs to be included in our examination of the different licensing strategies.

We shall show firstly to what extent the non-classical strategies are in fact forms of quasi-internalisation (section 7.1), and secondly, why they are indeed characterised by transaction costs that fall between the two extremes (internalisation and externalisation) (section 7.2). A new taxonomy of licensor–licensee relationships may then be defined.

7.1 NON-CLASSICAL STRATEGIES: INTERMEDIATE LEVEL BETWEEN INTERNALISATION AND EXTERNALISATION

It is appropriate to speak of externalisation only when the licensor hands over information and rights which enable his partner to integrate the process fully, either because a complete package is sold or because the licensee is able to combine the partial information received with what he already possesses. In this type of situation, the licensee is simply an acquirer of technologies and has no other relationship with the licensor.

However, it has already been pointed out that certain licensors have a different view of technology transfer. When they are pursuing a market or production strategy, they endeavour to retain control of part of the process involved. Consequently, the technology is *only partially externalised*. As a corollary, it may be said that they continue to internalise certain aspects of the technology:

- In a *market strategy*, the licensor retains the technology for manufacturing intermediate or complementary products, especially those that he feels to be of greatest 'strategic' importance (i.e., core technologies) and/or the most complex.
- In *production strategy*, the licensor may also retain 'strategic' technologies and be supplied by the licensee only with less vulnerable products. In any case, he controls all or part of his partner's outlets.

An analysis of this kind obviously means abandoning a monistic vision of technology. In reality, a firm possesses *not one but many* technologies that together contribute to the production of goods. It is rare for all of a firm's technologies to be transferred: usually only some of them are concerned, and because of the potential complementarity between these various technologies, quasi-internalisation may occur.

Seen from this standpoint, licensing has three main functions for a firm.

Extending the Firm's Scope

Transferring technologies may, indeed, be defined as a type of *coalition* that avoids the need to enlarge the firm, while at the same time securing additional resources. To use Porter's formulation, 'Coalitions are ways of broadening scope without broadening the firm'.[12]

In this respect, the analogy with franchising is very significant, as, according to Caves and Murphy,[13] it allows growth in an activity where the various stages of the manufacturing–marketing process have different economies of scale. Production has to be local and small-scale, whereas promotion has to be on a national or even international scale. The franchisor therefore hands over part of the process but retains, in particular, the function of promotion. In this way he broadens the *scope of his activity* via the intermediary of franchisees. The same way may be true in the licensing of technologies, if there are accompanying transactions between the partners, as the licensor acquires additional resources to complement his own and hence broaden the territory in which his firm is operative. The licensee is of course more autonomous than a fully-integrated workshop, but as he controls only part of the technology he is obliged to comply with the development strategy adopted by his partner.

Preserving the Firm's Power

The firm is sometimes presented as a place where power over the environment is sought. It is felt that the view taken here of the licensing relationship is more consistent with this idea than the traditional one of the licensor imposing restriction clauses in the contracts he signs.

Indeed, emphasis is laid here on motivations. It has been pointed out by Perroux, an outstanding French theoretician in the field of power in economics, who is unfortunately little known in Anglo-Saxon academic spheres, that once the idea of motivation is accepted, then the idea of power is a logical concomitant.[14] Motivation implies a project or plan and, ultimately, *the will to act on the environment*. In the traditional view of the licensing relationship, the licensor appears to have no project with respect to his partner; rather, he is simply concerned with obtaining sufficient remuneration and taking precautions. This is in some ways a simple restatement of the classical deal relationship between buyer and seller, with no interaction other than the onset of the contract and no attempt at creating a lopsided situation.

In many respects, however, technology licensing appears to correspond to the exercise of power, as has been shown by Jacquemin.[15] In this author's view, power lies in the *discretionary usage of surplus*, with the surplus being both the aim of the firm and the means of strengthening its power (expenditure on differentiation, dissuasion and concentration).

In the case of licensing, the exercise of power is expressed in a similar manner, for example by market strategies. The licensor uses his surplus, which in this case is technology – i.e., a technology that he possesses but does not need to use to the full – to increase his commercial stakes in an external market. In this case, the surplus is an intangible asset, but the procedure adopted is essentially the same. The transaction results in the *use of a surplus to reduce uncertainty*. There are obvious risks involved in developing the technology on an international scale; licensing technologies is one way of limiting such risks and uncertainties, while at the same time avoiding having to sacrifice commercial outlets, as these are guaranteed on a regular basis by the technologies themselves. It should be noted that the licensee is not necessarily inconvenienced by this exercise of power – the arrangement may indeed be to his advantage – but in all cases the local environment is subject to it.

However, if a classic formulation of licensing is adopted, then the analogy with Jacquemin's analysis breaks down to a certain extent, as in this case the granting of technologies is considered as a fairly opportunistic action which has no real connection with the firm's actual business. A power-based relationship is not excluded from this hypothesis, given the difference in capability of the partners, but the search for power itself is not motivated by the need to defend the organisation itself.

As a result of the above remarks, a distinction must now be made between quasi-internalisation and industrial cooperation. Market and production strategies might indeed appear to be particular types of industrial cooperation, as the partners render each other services in manufacturing and marketing products. However, although this comparison is reasonably justified, it should be pointed out that quasi-internalisation in fact refers to induced industrial cooperation that is a result of the partial granting of technologies.

Preserving Economic Rent

A firm with specific intangible assets will obviously tend to retain all the rent (excess profit) that can be obtained from the market as a result of those assets. The same is true on the international level and Vaitsos[16] has clearly shown the mechanisms used by multinationals to preserve their rents. Although this work concerned chiefly contracts between the parent company and its subsidiaries (i.e., internal transfers), most of these mechanisms are to be found in licensing agreements between independent firms.

Over and above the traditional restriction clauses, Vaitsos stresses in particular tying arrangements and their effects on prices. He explains that these clauses, which occur in as many as two-thirds of contracts, pave the way for overpricing, which he attempts to estimate.[17] In particular, he shows that tying arrangements are less strict in contracts between independent firms, and that overpricing is less pronounced.

It is extremely difficult to check these statements. However, it may be noted that, for many licensors, sales to the licensee are indeed the preferred channel for remuneration, even in the absence of tying arrangements (which are illegal in most countries). This is the typical compensation sought by those who are pursuing a market strategy. Conversely, with production strategy, it is possible to maintain a better competitive position on the home market and try to extract a rent there, using the licensee's less costly and/or better resources. Product flows between partners are therefore *effective back-up systems* for preserving rents.

Classical licensing contracts also enable licensors to try and obtain rents, but they are unable to avail themselves of the induced transaction channel and have to be content with the usual restriction clauses (limiting exports, compulsory notification of improvements, imposing technical assistance, etc.). Now, given the imperfect nature of the information market and, especially, the fact that neither partner is aware of the value of the information in question, the licensor has little chance of collecting the total rent. For example, if the technology should prove to be more profitable than expected, the licensor has virtually no way of revising his remuneration.

Here again, as in the two foregoing sections, it can be seen that market and production strategies form intermediate situations between internalisation, as described by Vaitsos, and the externalisation found in classic licensing strategies. To stress this point further, we shall now examine how quasi-internalisation is also seen in terms of *transaction costs*, this being of fundamental importance in choosing between internal organisation and resorting to the market.

7.2   QUASI-INTERNALISATION: INTERMEDIATE LEVELS OF TRANSACTION COSTS

Non-classical strategies for transferring technologies give rise to transaction costs situated between internalisation and pure licensing (technology strategy).

In support of this assertion, we shall adopt a two-stage approach. Firstly, we shall see that the conditions surrounding the transaction indeed fall between the two extreme situations. Secondly, we shall look in turn at each of the components of the transaction cost, in order to determine where the difference between quasi-internalisation and the two other types of transaction actually lies.

Transaction Conditions

According to Williamson, market failures that lead to internalising operations are due to the conditions in which transactions are performed. This author identifies two categories of condition that tend to generate high transaction costs, namely:

– Uncertainty combined with bounded rationality.
– Small number combined with opportunism.

These conditions arise in particular in technology transfers, as pointed out by Caves, Crookell and Killing,[18] as:

– The economic performance of a technology is always uncertain at the outset.
– Those involved in the transaction behave in accordance with the principle of bounded rationality, in that they consider a limited number of hypotheses in a sequential order.
– The number of potential licensors and licensees is limited – by definition in the case of the former, as their scarcity is the reason for the value of the technology and, in the case of the latter, because those competent to acquire the technology are also scarce.
– Opportunism or lack of candour and honesty in transactions is a

common type of behaviour, especially when suspicion exists between the partners, giving rise to a multiplicity of restriction clauses.

In short, conditions thus appear to be right for technologies to be transferred 'internally' rather than through the market. The internalisation theory is based on this observation.

However, it is argued that this analysis holds good only for certain licensing agreements, namely those which correspond to a technology strategy. In contrast, non-classical strategies are pursued in less favourable conditions and give rise to *lower transaction costs*.

There is thus uncertainty when the licensor is pursuing a market strategy, in that he retains full control of at least part of the process. This limits the risk involved for both partners, as there is less technological uncertainty. Furthermore, the licensor is thus able to stick to more familiar invoicing patterns. The uncertainty of commerical success on the host market is not done away with as a result, but complete internalisation does not necessarily offer any decisive advantages in this respect.

In a similar vein, there would appear to be less chance of opportunism with a market or production strategy. In such cases, as we have shown above, core technologies are retained by the licensor, while only those parts of the process involving the least risk are externalised. Moreover, he keeps a number of reliable cards up his sleeve in case the licensee should decide to adopt an opportunistic type of behaviour, as the licensee lacks several resources, and not only the technology. In comparison, internalisation (by setting up a subsidiary) implies even less opportunism as the 'internal' licensee is entirely dependent on the parent company for the technology, intermediate products, financing, management, etc.

With respect to opportunism and uncertainty, quasi-internalisation therefore constitutes an *intermediate situation*. The other two dimensions remain more or less unchanged in that bounded rationality is inherent in the transactions and there is no reason for the number of potential partners to be greater than in a classical licensing situation. It may be expected, in short, that transactions will be less complex and less costly than with complete externalisation.

In order to examine this point, we shall now review the various components of transaction costs.

Components of Transaction Costs

Signing a licensing agreement entails a number of expenses which may be assigned under five headings, in accordance with the work of Coase,[19] Buckley and Casson[20] and Teece:[21]

- The cost of *identifying a partner*.
- The cost of *negotiations*.
- The risk of *disclosure*.
- The cost of transferring *information*.
- The cost of *monitoring* the licensee.

Examination of each of these headings shows that the implementation of a market or production strategy entails a *lower* transaction cost than that associated with a traditional licensing contract. In the case of the first two items, in fact, the two types of strategy probably give rise to equivalent transaction costs; the difference is to be found in the last three.

The Cost of Identifying a Partner

When a technology is to be licensed fully, or externalised, the best possible partner must be found – that is to say, the one whose capabilities are most suited to the activity in view and who is thus most likely to reach a high level of performance.

In order to identify such a partner, information has to be collected from banks, clients, suppliers and competitors. The licensor also has to travel to meet the various possible candidates 'in the field'. All this entails a significant cost, which direct investment in a subsidiary would avoid.

On the other hand, in the case of quasi-internalisation, a comparable cost would be involved – since, even if the licensor is pursuing a market or production strategy, he needs a partner. There is no reason to assume that it would be systematically easier or more difficult to find a partner who would accept induced transactions (i.e., purchases or sales from or to the licensor) than in the case of a simple licence. On the one hand, it may be assumed that, as the technology transferred is less complete, more local firms would be capable of mastering it. On the other, as a closer relationship is proposed, it is possible that there would be fewer candidates willing to accept it. The outcome of these two conflicting trends will certainly depend on the local environment, and in particular on the average technological level of the economy. It is therefore our opinion that externalisation and quasi-internalisation are equivalent in this respect.

The Cost of Negotiations

Once the potential licensee(s) have been identified, it is then necessary to define reciprocal commitments and obligations with him/them in order to draw up a working agreement. This includes, in particular, the difficult question of evaluating the technology that is being licensed – that is to say, determining the benefits which it will provide for the acquiring party. We shall discuss this question later.

In addition, there is a whole series of mutual constraints on which the two parties must reach agreement, including guaranteeing exclusivity, the possibility of exporting, notifying improvements, etc.

Quasi-internalisation presents no particular advantages over externalisation with respect to these points. To a certain extent, the costs of negotiations are fixed. Moreover, though it may be assumed that the licensor has the means of retaliating against his partner and therefore needs to take fewer precautions, the complexity of the relationship, with its induced transactions, means that he has to negotiate additional points such as, for example, the increase in the degree of integration.

In short, as far as this point is concerned, there appears to be no significant difference between the two types of contract considered.

The similarity ends here, however. In the case of the last three points, quasi-internalisation enables certain economies to be made on transaction costs.

The Risk of Disclosure

We must now consider again the value of the technology being licensed, as this may entail considerable losses for the licensor.

Indeed, the technical know-how being handed over, either alone or with an authorisation, is of value only if kept secret. Now, the potential acquirer generally expects a certain amount of information to be disclosed in order to be convinced. This may in turn reduce the value of the technology if the party concerned acts in an opportunistic manner. This situation is referred to as the 'paradox of information'.[22] Information retains its value only when it is not disclosed to the acquiring party, but he needs the information in order to assess its value.[23] This problem is specific to information, and hence to the licensing of technologies.

Quasi-internalisation has an *undeniable advantage* over externalisation in this respect, as only a part of the 'information' is handed over and, as has already been stated, this does not include core technologies, for which the risk is greatest. If the information disclosed concerns only assembly techniques or basic components then the potential licensee has little to gain from behaving in a dishonest fashion, as the information received is not worth anything outside the context of the agreement. All else being equal, the implicit cost of disclosure is therefore lower in the case of a quasi-internalised transaction.

The Cost of Transferring Information

Once the contract has been signed, technical information has to be handed over to the licensee in order to implement the transfer. Information must therefore be 'codified' beforehand by the sender (in the form of drawings, computer software, etc.) and 'decodified' or interpreted by the receiver. In order for communications to function properly, those responsible for codifying and decodifying the information must have similar backgrounds, otherwise misunderstandings may occur; work would then have to be checked, and this would give rise to additional expenditure.[24] Teece's view,[25] which seems fully justified, is that these costs are a function of the difference in capability between the two partners and of the complexity of the technology in question.

Furthermore, the results obtained by Teece tend to confirm that internalisation enables these transfer costs to be less in comparison with the cost of transferring technology to an independent licensee.

Here again, the quasi-internalisation relationship appears to occupy a midway position as, with only part of the technology being handed over, *less information should be transferred*, thus entailing costs which are lower than in the case of complete externalisation, but probably higher than in the case of internalisation. This assumption is also made by Teece, who sees therein an economic justification for tying arrangements.[26] Without going so far, we may assert that quasi-internalisation helps to reduce transfer costs and that this may be a factor in negotiations between partners.

The Cost of Monitoring

Finally, the licensor must monitor the licensee's use of the technology, as regards both the quantity and quality of goods manufactured.

In the case of a simple licensing agreement, this is provided for in the contract clauses and will include regular inspections (possibly carried out by an auditor) to check the licensee's declarations, and regular examinations of production quality. These operations are obviously expensive and licensors try to make the licensee bear the corresponding costs, which is not always possible.

If the licensee is a subsidiary (as is the case with internalisation), the problem is posed differently. There is still a need for monitoring, but this is simpler to impose and there is no question of discussing who pays the cost.

The cost of monitoring in a quasi-internalisation context falls between these two. Since the licensee is independent, monitoring procedures and cost allocation have to be discussed. In contrast, monitoring will be *easier* and will provide savings in comparison with an externalisation solution. The licensee's activity can be estimated quantitatively by means of the deliveries made to him. The quality of manufactured goods is easier to assess, as the essential components – i.e., those which have a decisive effect on quality – continue to be produced by the licensor.

Similarly, with a buy-back provision in the agreement (in the case of production strategy) quality control is automatic. In short, quasi-internalisation does not obviate the need for monitoring but it provides *additional sources of information* on the partner's activities.

Finally, it may be concluded that quasi-internalisation provides *partial economies in terms of transaction costs*, so that it indeed constitutes an intermediate solution between direct investment and pure licensing. Moreover, it also provides economies with regard to the internalisation costs identified by Buckley and Casson[27] and Caves and Murphy.[28]

Indeed, while externalisation involves substantial transaction costs, this does not imply that there are no difficulties with internalisation. Buckley and Casson refuse to be dogmatic in this respect and state that an intermediate product market will be internalised if and only if the benefit (from internalisation) is greater than the cost involved.[29] We shall therefore quickly review the components of this cost in order to ascertain the implications of quasi-internalisation.

– The *first difficulty* to be faced in internalisation is the risk of *sub-optimality*, that is to say that the various technologies used at the same time in any given form of production function will have a different *minimum scale efficiency* (MSE). An integrated subsidiary is likely to exploit certain technologies below their profitability threshold. A quasi-internalisation strategy will enable technologies whose MSE are incompatible – and only those – to be externalised in the local environment.

- The *second difficulty* is the cost of coordination between subsidiaries and parent company. This may be very high, as the information system required to ensure adequate management and supervision may become very complex. Quasi-internalisation, in contrast, results in decentralisation, thus reducing management and supervision costs in a similar manner to franchising agreements, as the manager, who is in a position to take all necessary decisions, is 'on the spot'.[30]

- The *third difficulty*, which is specific to internalisation on the international scale, is the risk of political interference by the local authorities in the activities of the subsidiary, or even expropriation. This risk also exists when the technology is externalised, but in that case it is the licensee who is concerned. The same is true with quasi-internalisation, except that the licensee (who does not possess the complete technology) should continue with the relationship regardless of changes on the political scene. Quasi-internalisation therefore partially eliminates this risk.

Quasi-internalisation, in short, enables some of the costs of both internalisation and externalisation to be avoided. It is thus indeed an intermediate form of transaction for exploiting an intangible asset.

In conclusion, it should be stressed that the quasi-internalisation desired by a licensor becomes effective only if the licensee agrees. It is obvious that an agreement is the result of *convergence between two*

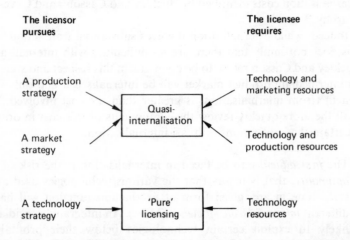

Figure 7.1   Type of licensing agreements related to partners' strategies

points of view. We should therefore examine how these two points of view combine to produce the different types of licensing agreement. This is shown diagrammatically in Figure 7.1.

This analysis may be validated in several ways. The technology pricing question would seem to be a particularly favourable field. Indeed, given the great differences in licensing strategies, it is certain that there will also be great differences in the manner in which they are implemented, and these differences are likely to affect various aspects of the question. One major example is the pricing policy, where a certain amount of confusion reigns. We feel that this new approach to the licensor–licensee relationship helps to sort out this confusion to some extent. The third part of this work is devoted to this problem.

Part III
Pricing Policies for Licensing Strategies

We have shown that licensing strategy is sometimes irrevocably linked to corporate strategy, and that technology is often licensed only in order to contribute to improving the real business of the firm, which is usually to sell products (finished or not) rather than information. This no doubt explains why some firms are more at ease negotiating *commercial transactions* than technology payments. The result, encountered fairly frequently where technology transfer to developing countries is concerned, is that technology payments are fixed to cover only the cost of the transfer.[1] The final outcome of this is to *open up markets* for the principal activity of the licensor, for his *main business*, rather than optimising the valorisation of the technology, which, in fact, the firm is not able to do. On this level, the accent is on profit margins earned through commercial transactions, rather than on royalty payments, and this indicates the desire to continue to pursue the normal activity of the firm, using its human and material resources.

This, moreover, is not limited to the transfer of technology to developing countries, but constitutes one of the main characteristics of licensing. Thus, having carried out detailed research on licensing agreements within the United States, E. B. Lovell concludes: 'The nature and purpose of the license grant is clearly a factor influencing the amount of the royalty payment, inasmuch as many companies report that they have both given and accepted some licences on a royalty-free basis, or in exchange for other licence rights'.[2] It is, in fact, in theory possible to show that the licensor can increase his total revenue by increasing the profit margins on his sales to the licensee, even if at the same time he relinquishes his royalties.[3]

We shall now see in reality how licensing strategies and technology pricing are linked. This will be the object of this third and final part of our work. From a practical point of view, we shall try first of all to confirm our idea concerning pricing policies (Chapter 8). With this aim in view we shall show that there is indeed a link between the *strategy followed by the licensor* and the *remuneration paid in return for the use of technology*. We shall study this link to show that there is no single formula for the fixing of technology prices, but that, on the contrary, the licensor adopts a pricing policy which varies according to the reasons for his granting a licence.

The results obtained will then permit us to *recommend* – to both licensors and licensees – *some of our conclusions*. In other words, we shall try to learn something from our observations. We shall thus be in a position to propose different examples of pricing policy (Chapter 9). These examples will be of use both to licensors and licensees, whose own concerns are taken into account. But above all, unlike the methods of price calculation available up to now, our examples are given *in relation to* the various licensing strategies. Negotiators should find in these examples information which will help them to establish their own line of conduct in discussing prices.

It should, however, be borne in mind that as we have not limited ourselves to a mere observation of methods used in practice: our recommendations are *not mere generalities*. This would lead to the more usual practices being transformed (quite unjustifiably in our opinion) into standard models. On the contrary, we also recommend lines of conduct which appear justified in theory even if they are still little used in practice.

8 An Empirical Analysis of Technology Pricing

The aim of the following pages is to present the results of the statistical work we carried out in order to test the validity of our approach. The fundamental aim is to ascertain whether or not technology pricing is indeed influenced by the strategy followed by the licensor. In other words, the importance of licensing strategies in technology pricing had to be evaluated. Naturally, a number of concepts had to be translated into measurable variables, in order to progress from a problematical situation to an empirical investigation.

Before reporting therefore on the results of our survey concerning the price–strategy relationship, Sections 8.2 and 8.3, we feel it necessary to explain our methodology and the limits imposed on collection of data.

8.1 METHODOLOGY

Starting from our technology pricing–licensor strategy assumption, we are going to formulate in more detail the type of results sought and the methods used to obtain them. We shall then proceed to give indications concerning data collection and conclusions reached.

The General Hypotheses

It is now evident, from what has already been written, that our proposition leads to the postulation of the existence of a causal relationship between the strategy characterising the agreement, and technology pricing.

If we consider that the licensor does not always wish to obtain maximum valorisation of his technological asset, for strategic reasons, we should note that the technology price is lower when the licensor follows a market or a production strategy as opposed to a technology strategy. However, this should not mean that what is lost in terms of technology payments is automatically recovered in revenues gained from transactions resulting from the agreement.

To try to validate these hypotheses, two types of investigation seemed possible. The first was to find a relationship between the technology price level and the type of strategy used. This meant in fact comparing the mean 'price' in contracts characterised by each of the generic strategies. But this analysis is not sufficient, in that it indicates only the existence of a relationship, the form of which remains to be shown.

The second type of investigation attempts to go beyond this limit, since its aim is to try to *recognise the form* of the price–strategy relationship. More precisely, it is trying to identify and measure the contribution of the various factors influencing the technology price. This means, of course, taking into account all the factors mentioned in previous publications on this subject, and moreover, testing new variables. Thus it is not merely a question of completely reviewing the method of calculating the technology price, but also of seeing the role played by strategic choices in the calculation of the price.

These two investigations will necessitate the definition of more precise hypotheses, and the use of the corresponding variables. Rather than proceed with this here, we prefer to leave it until the results are given (p. 141–59 below). Although some hypotheses and variables are used in both cases, others are used only in one of the two investigations.

Beforehand, however, some details concerning the collection of data for these investigations must be given.

The Data

To test the validity of our approach, we used information from two sources: first, licensors and secondly, licensees. It should be emphasised that the data collected concerns contracts effectively signed and not projects reflecting the intentions of one or other of the partners. Thus the costs and income figures on which our study is based are estimates given by the people interviewed, concerning amounts either already paid or which will be paid, over the whole duration of the contract.

It has not been possible to use exactly the same variables in both cases. It has, indeed, already been pointed out that the licensee does not know the exact cost of the transfer borne by his partner, and the licensor can only roughly evaluate the profitability brought by his technology to the licensee.

A certain amount of information, however, is known to both partners, such as the payments made by the licensee to the licensor, or the value of the commercial transactions made between them following the technology transfer. Even if the amounts of these variables differ slightly, at least they exist. It would be interesting to compare the results obtained from these two samples to see if they are consistent. If this were so, it is evident that the validity of the hypotheses would be more credible.

The fact that we did not interview both partners of an agreement but, on the contrary, licensors and licensees having no link in common (for evident reasons of confidentiality) does not seem to us to be of importance. What is important for our demonstration is that the same tendencies are confirmed in both samples, irrespective of the status of those interviewed (buyers or sellers). Unfortunately, this comparative approach was possible only for the first type of investigation – i.e., concerning the existence of a relationship.

8.2   THE EXISTENCE OF A PRICE—STRATEGY RELATIONSHIP

As stated previously, we have chosen to test the existence of a relationship between the technology price and the licensor's strategy, by comparing various contracts in each type of strategy.

Before examining the results, it is necessary to note our hypotheses concerning what is expected from such comparisons, and to define the variables which have been used.

Hypotheses Concerning the Existence of a Relationship

The working assumption mentioned at the beginning of this chapter can be expressed in several testable hypotheses:

- The *technology price differs* according to the strategy adopted by the seller, and accepted by or imposed on the buyer.
- The *technology price* is less when the technology transfer is accompanied by induced transactions (purchases or sales) forming the principal object of the agreement.
- Such transactions are *more frequent and financially more important* when the licensor follows a market or a production strategy.

– The technology payments may in some cases be *exceeded by income from induced transactions*, without there having necessarily been compensation.

The Variables

Taking into consideration the terms used in the above hypotheses, three sets of variables are required: the technology price, the strategy used, and the commercial transactions.

Variables Relating to the Technology Price

The idea of a technology 'price' raises certain difficulties when translated into a variable.

From previous research mentioned above, in particular research by Contractor and UNIDO, it can be seen that there are three possible types of variables: the technology payments, the multiple of the transfer cost and the licensor's share in the licensee's profit (*LSEP*).

1. Technology Payments We know that the various types of payment form a 'package', within which some compensation may be included.[1] Thus, the royalties do not form a satisfactory basis for comparison inasmuch as some licensors (and also some licensees) refuse to resort to royalties for various reasons. We therefore consider, as does Contractor,[2] that the unit of measure of the price can be defined as being the sum of technology payments over the duration of the agreement.

However, our approach differs here from that of Contractor, as we count as technology payments only those which explicitly concern the technology in question – i.e.:

– Lump sum and front-end payments.
– Royalties.
– Technical assistance fees.
– Training fees.
– Stock and dividend remittances, against industrial contribution.
– Technical compensation (communication relating to improvements, cross-licensing, etc.).

We have excluded from these technology payments the profits realised on commercial transactions resulting from the agreement (purchases or sales between partners). These profits are, however, included under the heading 'profit margin', as they do not actually result directly from the technology transfer.

2. *Transfer Cost Multiple* This second variable is complementary to the preceding one, as 'technology payments' represent an absolute value in money terms. It is therefore difficult to compare, as licence contracts can represent large sums of money without the price being particularly high. We therefore need a scale to fix the relativity of the technology payments. Contractor proposes comparing them to the 'transfer cost' – i.e., to the sum of the costs borne by the licensor when effecting a 'technology transfer'. He also gives a plan with five headings which we have reproduced in full: technical costs, legal costs, marketing costs, travelling costs, other costs.[3] Details of these five headings are given in Table 8.1.

The ratio between the two values – i.e., 'technology payments' and 'transfer costs' – gives a multiplier coefficient of the profitability of the agreement, and therefore constitutes a particularly interesting relative price indicator.

On this point, our approach differs again from Contractor's. We recognise, as he does, the existence of a time lapse between the payments and the costs, which theoretically makes it necessary to apply a discounting factor to both series, in order to make them homogeneous. However, for practical reasons (the complexity of information received from year to year) which would have been detrimental to the wide sample studied, we have preferred to work on gross values. This concession to practicality does not appear to have had too great an influence on results, as Contractor's results calculated on gross values are very near to those calculated on discounted values.[4]

3. *The Licensor's Share in the Licensee's Profit* One last variable, created by UNIDO, presents a certain analytical interest. This is the consideration not only of royalties but of all technology payments when calculating the share retained by the licensor from the profits accrued to the licensee through using the technology. This type of calculation implies the collection of data concerning the profitability of the technology in question, which can be effected only through the licensees. Unfortunately, most of them do not keep records per 'project', and it was therefore impossible to assess, to any precise degree, the price of technology using this variable.

Table 8.1   Transfer costs

Technical costs	– Cost of drawing up and preparing blueprints, dies, models and other technical studies
	– Engineers' and technicians' salaries and living expenses
	– Cost of training the partner's technical personnel
	– Cost of technology adaptation
	– Cost of construction and installation
	– Auditing and quality control costs
	– etc.
Legal costs	– Cost of obtaining patents and registration of trade marks
	– Cost of legal consultants, court cases
	– Cost of negotiating and preparing final agreement
	– etc.
Marketing costs	– Market survey costs
	– Training partner's sales personnel
	– Advertising and publicity
	– Salaries and living expenses of sales personnel lent to the licensee
Travel	– Travelling and living expenses of managers and negotiators
Other	– Costs not mentioned above, in particular: costs incurred previous to the negotiation but nevertheless directly linked to the agreement (e.g., prospecting costs)
	– Cost of investment in the equity of the partner company (minority joint venture)

Variables Relative to the Type of Strategy Used

On the other hand, the concept of strategy does not pose any special problem when translated into variables, since it appears in three Generic Strategies (technology, market and production). Each of the people interviewed put the licence under study into one of these categories, indicating at the same time the underlying reasons for the contract.

Variables Relative to Commercial Transactions

Three variables affect the strategy followed.

1. Induced Transactions   These are transactions resulting from the licensing agreement. Several types of transactions are possible between partners: the sale of equipment, the sale of components and spare parts, the sale of goods complementary to those already on the market, the purchase of manufactured goods and the purchase of intermediate goods. These transactions are from the licensor's point of view – symmetrical transactions naturally exist from that of the licensee. The number of types of transactions effected between the two partners depends on the nature of the relationship resulting from the licensing agreement.

2. *Bilateral Transaction Flows* In addition to a mere list of transactions, it is useful to have an estimate of the financial sum represented by purchases and sales over the contract period. The relative importance of these transaction flows in relation to technology payments will help to decide the strategy to be followed.

3. *The net profit margins on purchases and sales* These represent the last variable,[5] the absolute value of the profits estimated by the licensor on transactions he has made resulting from the licensing agreement. Here again, the comparison of this variable with the technology payments will be interesting as far as the real motivation of the licensor's agreement is concerned.

The Evidence for the Relationship

As mentioned earlier, whenever possible we shall present the results of the two licensor and licensee samples. This is the case with several variables. We shall examine here the results of both samples, using the same method for both. This consists in finding a strategy–price relationship by splitting the sample into sub-samples according to the strategy used, and by comparing the level of variables of prices used.

Strategy–Price Relationship in the 'Licensor' Sample

It was possible to form only two sub-samples among the licensors, as only one contract fell into the 'production strategy' category; 20 contracts came under the heading 'technology strategy' and 12 under 'market strategy'. We have therefore made only two sub-samples; one 'technology' and the other 'market and production', which groups together all the contracts characterised by non-classical strategies.

Table 8.2 shows the arithmetical means obtained by our selected variables.

It is clear that different strategies are represented by different payments. It can be seen, first of all, that market and production strategies are usually related to contracts of *lesser financial importance*. This is not because of lower transfer prices: on the contrary. The content of the transfer seems heavier – i.e., more important services are accorded to the licensee by a licensor who uses a market strategy. Compared to agreements made in the 'technology' strategy, this type of transfer necessitates more work in the field of technical surveys, training and technical assistance, etc. so that the mean 'multiple' is considerably less in non-classical strategies. This means that the compensation for the technology granted – *its 'price'*, according to one of the meanings we have adopted – *is smaller*, as assumed by our second hypothesis.

If we consider the profits created by commercial transactions, the analysis can be taken further, and it can be seen that these profits are considerably less when technology strategy is used than when market or production strategies are used.

This does not, however, mean that *strict compensation* should intervene between these two payment headings. If the ratio of profits plus technology payments to transfer payments is calculated, it can be seen that this multiple (Table 8.2, line 6) remains lower in the case of market and production strategies. This is in accordance with the fourth hypothesis.

One can therefore conclude that the choice of a licensing strategy *does have some influence* on the technology payments.

It is interesting at this point to compare these results, obtained from the 'licensor' sample, to those obtained from the 'licensee' sample.

Strategy–Price Relationship in the 'Licensee' Sample

Table 8.3 gives support to some of the conclusions reached concerning the licensor sample.

Technology payments are, on average, lower where market and production strategies are used. Although the very limited character of this variable at this level of analysis has been stressed, it is interesting to note that in the licensee sample also, licensing agreements cover more or less large sums according to the strategy used. It is also remarkable that the amounts are *approximately the same*, since the technology payments are about twice as large where technology strategy is used as

Table 8.2 Arithmetical mean of price variables and transactions according to the strategy followed in the sample of licensors (33 contracts)

	Market and production strategy (13 observations)	Technology strategy (20 observations)	Whole sample (33 observations)
1. Technology payments (1000 francs)	37 504 (79 029)	65 508 (227 211)	53 764 (181 069)
2. Transfer costs (1000 francs)	30 439 (92 344)	17 103 (66 809)	22 521 (78 468)
3. Transfer cost multiple	3.65 (4.67)	202.31 (749.83)	116.22 (572.98)
4. Profit margins on commercial transactions (1000 francs)	96 783.0 (186 465)	1068.9 (3505.9)	42 545 (131 617)
5. Profit margin/technology payments ratio	526.16 (1655.1)	0.28 (0.68)	214.53 (1087.6)
6. Whole multiple of the transfer cost (1000 francs)	19.80 (35.40)	217.25 (770.56)	128.74 (581.20)
7. Bilateral transaction flows (1000 francs)	660 769 (1 836 389)	57 065 (228 718)	310 231 (1 238 273)

Notes:
1. The figures in brackets represent the standard deviations corresponding to each arithmetical mean.
2. The number of 'useful' observations changes from one variable to another, so that the third column is not necessarily the weighted mean by 13 of the first column and by 20 of the second column.
3. Note that line 3 of the table indicates the mean ratios of technology payments/transfer costs obtained from each contract, which is not the same as the ratio of mean technology payments to mean transfer costs. This also applies to line 5 of the table.

Table 8.3 Arithmetical mean of price variables and transactions according to the type of strategy used in the licensee sample (1000 francs)

	Market strategy (11 observations)	Production strategy (6 observations)	Technology strategy (12 observations)	Whole sample (29 observations)
1. Technology payments	6156.50 (10 633.0)	956.67 (979.89)	13 383 (19 187)	7881.3 (14 580.0)
2. Sales (to licensor)	735.45 (1876.40)	21 945.0 (29 708.0)	0	4991.4 (16 401.0)
3. Purchases (from licensor)	42 272.0 (55 953.0)	12 500.0 (10 530.0)	4640 (11 353.0)	21 108.0 (40 034.0)
4. Bilateral transaction flow	43 007.0 (55 791.0)	34 445.0 (39 898.0)	4640 (11 353.0)	26 100.0 (43 849.0)

Notes:
1. The figures in brackets represent the standard deviations.
2. The number of effective observations changes from one variable to another, so that column 4 cannot be calculated directly as a function of the other three.
3. Line 4 represents the mean bilateral transaction flow, and is therefore not equal to the sum of mean sales plus purchases.
4. Contracts were placed in one of the three strategies by the licensees themselves, according to their idea of the motives of their technology supplier.

where market strategy is used, both in the licensor and in the licensee sample. This implies that the technology is 'cheaper' in the second case than in the first.

It is possible to check the relationship already noted between the strategy followed and the commercial transactions (purchases and sales) resulting from the licensing agreement. The mean amount of bilateral transaction flows is much greater when the licensing agreement motivation is market- or production-oriented. Allowing for a profit margin of 15 per cent, which would appear reasonable from our observations, the profits in this sample also *are greater than* the actual technology payments where contracts are made using 'market strategy'.

It therefore appears certain that a relationship exists between strategy and technology pricing. Data on profits made by the licensor (*LSEP* in UNIDO terminology) would permit an even more detailed analysis. But, as we have already explained, this data does not exist, even with the licensees. The rare information we were able to obtain did not enable us to come to any general conclusion, as it covered only 6 contracts. We shall therefore mention it only in passing, disconcerting as it may be! If the ratio of technology payments to profits earned through the acquisition of technology is calculated, it can be seen that this ratio is on average 6.9 per cent in contracts characterised by market or production strategy (3 contracts) as against 77.3 per cent for those based on technology strategy (3 contracts). If these results were confirmed on broader sampling, it would be legitimate to assume that technology sold using technology strategy does indeed have a higher price.

It appears possible after this first investigation that strategy and technology prices are related, and the results shown here support the four hypotheses put forward. It is the form taken by this relationship which we will now examine.

8.3 FORM OF THE STRATEGY–PRICE RELATIONSHIP

We have chosen to study the form of this relationship using the multiple regression technique. This is ideally suited to our project as it makes it possible to determine which variables have the most influence on technology pricing. But this investigation could be feasible only concerning the 'licensor' sample, which provided more data.

Having introduced the hypotheses formulated at this stage of the analysis, we shall describe the variables which were the object of the statistical processing, and present the results of the regression which show the variables having the most influence on pricing. With this aim in view we shall split the sample according to the type of strategy pursued; this will introduce a new dimension in technology pricing.

Hypotheses on the Form of the Relationship

The main hypothesis here is that the way by which the price is fixed depends on the strategy pursued, and that – contrary to what is maintained by other researchers – there is no *one formula* for pricing, applicable to all technology transfers, but *several formulae*, each suited to a different type of agreement.

This analysis can be transposed into more detailed hypotheses; it implies that:

- Pricing is influenced differently by determining factors (different regression coefficients) according to the *type of strategy* used.
- The *determining factors are not the same*, some having a significant influence in the case of one strategy but not in the case of another.
- The statistical quality of the relationship is improved if it is considered *per type of strategy*, and not for all strategies together.

Even then, the determining factors in question must first be identified. For this, we have used both research previously carried out and logical reasoning within our own analysis.

The investigation reported here is thus not intended to invalidate previous research, but to complete it by taking strategy into account. It is with this aim in view that we are going to enumerate the multiple variables which influence pricing.

The Variables

We shall, of course, distinguish between variables which are expressions of pricing and those considered at first sight to be determinants of pricing (see Table 8.4).

We have kept five of the first type (referred to as dependent variables). In fact, only two have given any significant relationship from a statistical point of view. These are:

- *Technology payments* – i.e., the sum of payments accruing to the licensor, correlated with the execution of the contract.

Table 8.4   Variables used in the regression analysis   151

Variables	*Measures*
Dependent variables	
– Technology payments	1000 FF
– Technology net returns	1000 FF
– Transfer cost multiple	unit
– Whole multiple of transfer cost	unit
– Royalties	% (basic rate)
Independent variables	
– Transfer cost	1000 FF
– Licensor's stage of technology development	1. Design 2. Prototype 3. Pre-series and market test 4. Launch 5. Full operation 6. Decline 7. Discontinued operation
– Research and development expenditure by licensor	% of sales
– Market position of licensee	1. Leader 2. Challenger 3. Marginal 4. Beginner
– Licensee status	1. Private firm 2. Public corporation
– Location of licensee	1. LDC and NIC[1] 2. CPE[2] 3. DC[3]
– Number of induced transactions	0–5 according to typology
– Sales on induced transactions	1000 FF
– Operating profit on induced transactions	1000 FF
– Relative weight of agreement (ratio of technology payments to licensor annual sales)	unit
– Licensor size: annual turnover at time of signing agreements	1000 FF

Note:
1. Less Developed Countries and Newly Industrialised Countries
2. Country with a Planned Economy
3. Developed Countries

- *Net technology payments* – i.e., payments remaining after the deduction of transfer costs. In other words, the operating profit made by the licensor thanks to the difference between income and costs directly linked to the licensing agreement. This is the closest we have been able to come to expressing pricing.

We shall also, however, mention another variable which has already been defined: the transfer cost multiple.

We should like to explain briefly here the type and influence that the independent variables which were selected have generally on the pricing process:

- The *transfer cost*, which according to Contractor should be positively linked to the price.
- The *stage of development of the technology*, which should be inversely linked to the price (the older the technology, the lower the price).
- *Research and Development expenditure*, which is often considered as being an indication of the cost of developing the technology, which itself determines the price level. Here there will therefore be a positive regression coefficient.
- *Licensee's profitability*, that we can estimate only by studying his position on the market. This variable is calculated on the basis (confirmed by many researchers, including the PIMS programme) that the greater the market share, the higher the profit. As the measure we have used is inversely linked to the market size, we can assume an inverse relationship with technology price.
- The *licensee status* (private or public) assuming that the latter have more negotiating power, and could therefore pay a lower price, as it is sometimes admitted.
- The *geographical location of the licensee*, which can affect the negotiating power.
- The *number of induced transactions* which, like the two following variables, should be linked to the price by a negative regression coefficient (the greater the number of transactions, the lower the technology price).
- *Turnover* on induced transactions.
- *Operating profits* on induced transactions.
- The *relative economic weight of the agreement for the licensor* (calculated by the ratio of technology payments to his annual turnover), which should be positively linked to the price, in as far as a high ratio could indicate that the technology transfer is no longer a marginal activity.

– The *size of the licensor*, calculated by his annual sales at the time of the agreement and which should obviously be positively linked to the price, on account of the effect of size (large firms, large operations).

The types of relationship that we have dealt with are relative to a global approach to licensing agreements. They are, so to speak, the 'classical' determinants of technology pricing. We shall therefore not be content merely to test them, but we shall try to see how they react when considering the licensor's strategy. We expect that the influence they exert on pricing will *differ* from one strategy to another.

Price Determinants

Although we have chosen a very limited number of variables representing the pricing, none of which touch on the share retained by the licensor in the profit made by the licensee (*LSEP*), two gave no significant results. We can therefore work on the first three: technology payments, net technology returns, and transfer cost multiple. On this last variable there is unfortunately little to say.

This price variable is interesting on account of its relative character, and it would be valuable to know why some licences enable the licensor to benefit from a wider margin on the transfer cost compared to others. We therefore tested a regression, the results of which are, it must be admitted, unconvincing: weak R^2 values, weak t values. At the best, only one variable has a satisfactory t: the ratio of technology payments to turnover – i.e., the variable giving the level of dependence of the firm on the technology transfer. But it is evident that no conclusion can be drawn from such a result. In fact, the absence of a conclusion on this point should be stressed.

On the other hand, we obtained quite interesting results concerning technology payments and net technology returns. We shall first of all present those covering the whole sample, and then those specific to the two sub-samples, which we reached by putting together all the agreements coming under a market strategy (12 observations) and all those under a technology strategy (20 observations). As we had only one 'licensor' agreement coming under the production strategy heading, we ignored it. This does not seriously affect our demonstration as the confrontation of the two remaining sub-samples should be sufficient to analyse the effect of different strategies on pricing.

Results on Whole Sample

We have tried, at this level, to identify the determining variables using those which could possibly play a role. After each regression we eliminated those which had the lowest explaining power (the smallest t values). Using this process, we were thus able to see which variables contributed most to the price calculation. These results are given in Table 8.5, and in equations 10.1, 11.1 and 12.1 of Table 8.7 concerning technology payments, and Table 8.6 and in equations 14.1, 15.1, and 16.1 of Table 8.8 concerning net technology returns. Similar conclusions are reached with both these price variables.

It can be seen that 'classic' factors such as the market position of the licensee, or the size of the licensor calculated on annual sales, count *very little*. It is true that the first equations ((2) and (6)) give highly significant results statistically, but that is due partly to the small amount of data actually used. In fact, the software used automatically eliminates any examples where data concerning the variables is incomplete.[6] Thus, the more variables there are, the greater the risk of having only a small sample which is actually processed.

It can thus be seen that only two variables appear significant: *transfer costs* and *profits* resulting from commercial transactions. The mathematical sign for the coefficients corresponds exactly to the hypotheses expressed: positive for the transfer costs (the higher the costs, the 'more expensive' the technology) and negative for the profits on commercial transactions (the higher the profits, the lower the technology price). These facts are valid for almost all the equations, including those shown in Tables 8.7 and 8.8.

On the other hand, *several variables* must be *rejected*. Apart from those already mentioned, this concerns variables connected with the development stage of the technology in question, the status of the licensee (private or public), the intensity of the commercial relationship with the licensee (additional turnover and number of transactions) and the intensity of Research and Development with the licensor (R & D expenditure).[7]

We can therefore partly recognise earlier results, namely Contractor's; but only partly, as we have brought a new element to light: the operating profits made by the licensor thanks to transactions resulting from the agreement. This last element is of a different type in that it introduces a new dimension in the various payments made to the licensor.

However, if we cease to consider that the licensor population is

Table 8.5 Results of regression analysis on 'technology payments' variable (whole sample)

Equation no.	Dependent variable	Constant	Transfer cost	Operating profit on transactions	R&D expenditure by licensor	No. of transactions	Licensee status	Development stage of technology	Additional turnover	Market position of licensee	Licensee's turnover	F-values	Degrees of freedom	$\bar{R}^2$	R^2
										Independent variables					
(2)	Technology payments	-22457 (-1.48)	3.27 (28.25)	-0.12 (-0.96)	-282.09 (-3.98)	-6178.1 (-3.42)	-8304.5 (-2.68)	9538.5 (3.10)	0.024 (0.86)	57.01 (0.03)	0.00008 (0.81)	3157.9	11	0.9996	0.9993
(3)	Technology payments	15311 (0.79)	3.36 (24.42)	0.055 (0.31)	-172.83 (-0.34)	-3893.5 (-2.55)	-5339.6 (-1.30)	-146.70 (-0.04)	0.0007 (0.02)			2722.1	14	0.9993	0.9989
(4)	Technology payments	49935 (1.49)	3.36 (22.06)	-1.23 (-14.69)	128.71 (0.09)	-5695.0 (-0.68)	-23050 (-1.09)					142.12	17	0.9766	0.9668
(5)	Technology payments	9127.9 (0.19)	2.98 (12.00)	-1.07 (-7.07)	363.64 (0.46)	17720 (1.03)						39.88	19	0.8936	0.8712

Note:

The figures in brackets represent Student *t*-values obtained by the variable for the considered regression. The regressions presented here have been computed by the method OLS, to which was applied automatically a process to reduce the autocorrelation of the residues. Only equations without autocorrelation were retained.

Table 8.6 Results of regression analysis on 'technology net returns' variable (whole sample)

Equation no.	Dependent variable	Constant	Transfer cost	Operating profit on transactions	R&D expenditures by licensor	No. of transactions	Licensee status	Development stage of technology	Additional annual turnover	Market position of licensee	Annual turnover of licensor	F-value	Degrees of freedom	$\bar{R}^2$	R^2
(6)	Technology net returns	−22713 (−1.49)	2.275 (19.62)	−0.115 (−0.95)	−283.08 (−4.00)	−6223.9 (−3.45)	−8217.1 (−2.65)	9564.10 (3.10)	0.024 (0.85)	95.58 (0.05)	0.00008	1543.7	11	0.9992	0.9986
(7)	Technology net returns	15317.0 (0.79)	2.357 (17.16)	0.055 (0.31)	−173.02 (−0.34)	−3892.5 (−2.55)	−5345.5 (−1.31)	−147.03 (−0.04)	0.0006 (0.02)			1333.0	14	0.9985	0.9978
(8)	Technology net returns	49944.0 (1.49)	2.358 (15.69)	−1.231 (−14.69)	127.90 (0.09)	−5691.3 (−0.68)	−23067.0 (−1.09)					66.425	17	0.9513	0.9370
(9)	Technology net returns	9055.8 (0.18)	1.982 (7.98)	−1.066 (7.07)	363.02 (0.46)	17737.0 (1.03)						17.03	19	0.7819	0.7360

Note:
The figures in brackets represent Student t-values obtained by the variable for the considered regression. The regressions presented here have been computed by the method OLS, to which was applied automatically a process to reduce the autocorrelation of the residues. Only equations without autocorrelation were retained.

homogeneous, and if the sample is split according to the strategy followed, other dimensions appear.

Results on Sub-samples

Tables 8.7 and 8.8 enable us to compare the price factors in various situations: with a market strategy, a technology strategy and without any strategy – i.e., using all the samples together as a whole. Several equations were tested, on a basis of four determining variables, both for technology payments and for net technology payments.

If we compare the regression analysis of the sub-samples both with each other and with the sample as a whole, we can see that our three hypotheses are confirmed. It can be seen that, apart from a few exceptions, the same conclusions are reached concerning both the total price (technology payments) and net price (net technology returns). Thus in equations (12.1)–(12.3) on one hand, and (16.1)–(16.3) on the other, we can see that the form of the relationship strongly differs according to the type of strategy used.

First of all, the regression *coefficients* are *never the same*, and the signs are often contradictory. When a technology strategy is used, a higher transfer cost will imply a higher technology price (coefficient 3.125). On the contrary, when a market strategy is used, the relationship is reversed (coefficient -0.237). This is also significant when technology net returns are used as a dependent variable (the corresponding coefficients are 2.125 and -1.237 respectively).

Secondly, the *determining variables* which are significant in one equation, may not be so in the other. Thus if the transfer cost influences the technology payments in the 'technology' sub-sample (t-value = 13.73), it does not do so in the 'market' sub-sample (t-value 0.93). On the contrary, the margins on transactions which are significant in this latter sub-sample are not so in the other. This is also true concerning the other dependent variable.

Finally, it can be seen that the statistical quality of the correlations is usually significantly improved when the different strategies are taken into account (best R^2 and t), rather than taking all the contracts together as a whole.

Our main hypothesis concerning pricing can thus be emphasised – namely that the type of relationship, and its contents, are affected by the licensor's strategy:

Table 8.7 Results of regression analysis on 'technology payments' variable, with sample distribution

Equation no.	Dependent variable—technology payments	Constant	Transfer cost	Operating profit on transaction	Licensor's R&D expenditure	No. of transactions	F-value	Degrees of freedom	R^2	R^2
(10.1)	Whole sample	9127.9 (0.19)	2.98 (12.00)	-1.07 (-7.07)	363.64 (0.46)	17720 (1.03)	39.88	19	0.8936	0.8712
(10.2)	Market sample	-38715 (-1.14)	-0.467 (-1.59)	-0.566 (3.19)	1371.5 (0.36)	14168 (0.98)	32.15	6	0.9554	0.9257
(10.3)	Technology sample	8156.1 (2.45)	2.994 (15.38)	7.228 (1.79)	-347.14 (-0.71)	-440.1 (-1.89)	6065.9	8	0.9997	0.9995
(11.1)	Whole sample	41387 (1.24)	3.093 (13.90)	-1.071 (-6.98)	709.78 (0.85)		51.37	20	0.8851	0.8679
(11.2)	Market sample	-5465.5 (-0.54)	-0.332 (-1.25)	0.489 (3.08)	3437.3 (1.21)		40.76	7	0.9458	0.9226
(11.3)	Technology sample	2681.1 (0.92)	2.983 (13.36)	7.015 (1.53)	-242.49 (-0.92)		6204.8	9	0.9995	0.9994
(12.1)	Whole sample	25422 (1.76)	2.865 (12.45)	-0.908 (-6.38)			82.46	27	0.8593	0.8489
(12.2)	Market sample	-3553.4 (-0.44)	-0.237 (-0.93)	0.478 (3.89)			41.60	9	0.9024	0.8807
(12.3)	Technology sample	5889.1 (3.03)	3.125 (13.73)	3.571 (0.79)			5736.8	14	0.9988	0.9986

Note:
1. The notes to Table 8.5 also apply to this table.
2. Note that the estimate of regression coefficients in the two sub-samples has also been made taking care to avoid splitting and thus losing data, by introducing an indicative variable for the strategies. This made it possible to obtain results very similar to those mentioned, and we therefore have not presented them.

Table 8.8 Results of regression analysis on 'net technology returns' variable, with sample distribution

Equation no.	Dependent variable—technology payments	Constant	Transfer cost	Operating profit on transaction	Licensor's R&D expenditure	No. of transaction	F-value	Degrees of freedom	R^2	$\bar{R}^2$
(14.1)	Whole sample	9055.8 (0.18)	1.982 (7.98)	−1.066 (7.07)	363.02 (0.46)	17737.0 (1.03)	17.03	19	0.7819	0.7360
(14.2)	Market sample	−38715 (−1.14)	−1.467 (−5.00)	0.566 (3.19)	1371.6 (0.36)	14168.0 (0.98)	10.269	6	0.8725	0.7876
(14.3)	Technology sample	8159.10 (2.45)	1.994 (10.24)	0.7228 (1.79)	−347.36 (−0.71)	−4441.5 (−1.89)	2958.80	8	0.9993	0.9990
(15.1)	Whole sample	41354 (1.24)	2.093 (9.41)	−1.071 (−6.98)	709.43 (0.85)		21.65	20	0.765	0.729
(15.2)	Market sample	−5465.5 (−0.54)	−1.332 (−5.02)	0.489 (3.08)	3457.3 (1.21)		12.735	7	0.8452	0.7788
(15.3)	Technology sample	2661.4 (0.92)	1.982 (8.91)	7.032 (1.54)	−243.27 (−0.93)		3052.10	9	0.9990	0.9987
(16.1)	Whole sample	25421 (1.76)	1.865 (8.11)	−0.908 (−6.38)			33.167	27	0.7107	0.6893
(16.2)	Market sample	−3553.4 (−0.44)	−1.237 (−4.84)	0.478 (3.89)			11.501	9	0.7188	0.6563
(16.3)	Technology sample	5897.9 (3.03)	2.125 (9.33)	3.566 (0.79)			2790.7	14	0.9975	0.9971

Note:
1. The notes to Table 8.5 also apply to this table.
2. Note that the estimate of regression coefficients in the two sub-samples has also been made taking care to avoid splitting and thus losing data, by introducing an indicative variable for the strategies. This made it possible to obtain results very similar to those mentioned, and we therefore have not presented them.

- A *technology strategy* entails a price (technology payments and net technology returns) which is proportionally higher as the transfer cost increases.
- A *market strategy* entails prices, however defined, less than proportional to the profit margins on commercial transactions resulting from the transfer.

The meaning of these relationships is more important than the exact value of the coefficients concerned. This can be interpreted in terms of a *different logic*. When a technology strategy is used, the licensor is trying to obtain the highest possible price for the transfer of technology – i.e., the 'classic' behaviour of a seller. Even if there is no maximisation process, but only a desire to attain a satisfactory return, the aim of the transaction remains the technology. On the contrary, when a market strategy is used, with all the ensuing types of collaboration, the payment is affected by profits resulting from the sales of goods.

It is, moreover, interesting to note that, as we expected, the price equation in the technology sub-sample is relatively closer to the regression equation of the whole sample. This is in line with our remark concerning earlier studies, when we pointed out that these studies accepted as a generality what is really only the specific logic relative to certain technology transfers, made under a 'technology' strategy. On the whole, the results of this processing indicate the existence and the form of a licensing strategy–price relationship.

It can be considered that when a market strategy is followed, the price is lower, as shown by comparisons carried out on both licensor and licensee samples. The regression analysis results show, moreover, that pricing practices vary somewhat according to the type of strategy adopted: other factors intervene in addition to the price of the technology transfer, such as profit margins on transactions made as a result of market strategy.

It would appear that the pricing policy of licensors, which is agreed to or accepted by their licensees, is greatly influenced by the motivations which led to the agreement. There is not *one*, but *many* types of technology selling behaviour, and these are naturally characterised by the differences on one of their components: pricing policy.

We can now make recommendations, certain as we are that our view of pricing policy corresponds to a certain reality put into practice by firms. We can already depart from many pricing models which exist only in the imagination of their creators.

9 Internationalisation Strategies and Technology Pricing Policies

An important step will now consist in drawing conclusions from our research in the form of recommendations, for the licensee as much as for the licensor. We will do this on the basis of our own results, while also using those of previous writers on the subject, which we have analysed at length.

This chapter may appear repetitive in some ways. But it must be appreciated that it has a different status to those analytical chapters which came before. The aim is, in fact, to assemble all the knowledge which has been accumulated so that it can be put into practice by the agents. So we will necessarily have to go back to the essential variables of licensing strategy and pricing policy. But we will do so from the angle of their *application*, which will mean sometimes adopting an attitude of simplification.

A fundamental aspect of our demonstration was to relate the determination of price to the strategy being followed by the two partners. So our recommendations will go back to a stage prior to the determination of the price, to deal with the options of international development for the licensor and of technical development for the licensee.

We will devote the first part of this chapter to the licensor's strategic options for internationalisation, before going into his pricing policy in the case of an external transfer of technology, in the second part. The third part will give us the opportunity of suggesting to the licensee an approach for the strategic analysis of the options facing him.

9.1 ALTERNATIVES IN THE INTERNATIONALISATION OF THE FIRM: THE STRATEGIC CHOICE OF THE LICENSOR

It is necessary, first of all, to define once more the type of alternatives which a firm faces when it has taken the option of internationalisation.

It will be only in the second stage that we can study the factors which may be determinant in this choice.

The Elements of Choice

We must indeed come back to the forms of internationalisation after completing this research, because the traditional alternatives (exporting, direct foreign investment and licensing) did not seem satisfactory to us. The fact that the third element of choice covers widely different strategies has prompted us to add a further option: 'quasi-internalisation'.

We have already seen how licensing could be used to 'internalise' in part the application of the company's 'technological assets' by not transferring the whole of the process to the licensee. Through production and market strategy, the licensor retains control over the essential part of this technical knowledge. 'Dis-internalisation' (to quote Dunning's expression) effectively takes place only with the technology strategy which we will refer to as 'pure licensing'.

The company is faced with four possible options in carrying out its internationalisation:

- Exporting.
- Direct foreign investment.
- Quasi-internalisation.
- Pure licensing.

The second and third options include in fact two symmetrical possibilities, in each case: market strategy and production strategy.

All these distinctions have to be made before we can suggest a model for choosing between these different forms of internationalisation, which we will now do. But we must point out again that this model applies only to companies considering horizontal transfers, which were the object of our research, leaving out those companies which specialise in the production of technologies for external use.

The Determinants of Choice

Our ambition here is to suggest an analytical approach to help a company choose its forms of internationalisation. This approach – i.e., this model for formulating a strategy – is intended to be at the same time practical and objective: practical in the sense that it should be applied without any difficulty, and objective in the sense that it is not based simply on intuition (which is too often the case with this kind of approach), but essentially on factors which have been tested.

We will first present the framework of the model, then we will explain the action of the various factors on the choice of the option.

The Model for Choosing the Options

We believe that three aspects should be taken into consideration when choosing the form of internationalisation:

- The company's *current priorities*.
- The *target markets*.
- The *resources* available to the company.

This means that the form of internationalisation should be chosen in such a way as to make these three dimensions compatible. The way of achieving this is shown graphically in Figure 9.1.

The first step consists in defining the company's current priorities in order to determine whether, and in what way, internationalisation is a relevant strategic issue.

Then the second step can start, which entails an analysis of the chosen target markets, and of the resources available to the company to reach them.

This implies that the model will allow only the most efficient form of internationalisation to be determined, target-market by target-market – i.e., for national, regional or multinational markets depending on their degree of homogeneity.

It will therefore be necessary to examine each of the possible target markets one after the other. Each one will be analysed under three aspects (its commercial, industrial and political features) in order to determine which form of internationalisation is most appropriate.

Then we must study the forms of internationalisation which the company's resources will allow. These are divided into four constituent parts: marketing, technical, human and financial. The framework of the model having been defined, we can bring forward further details on each of its constituents.

1. The Company's Current Priorities   The choice of a form of internationalisation implies that the company clarifies its current priorities within its 'business strategy' (as opposed to its 'corporate strategy').

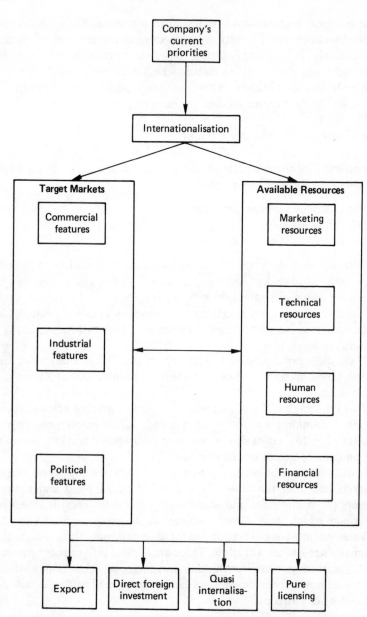

Figure 9.1   A model for choosing the form of internationalisation

By referring to what has been studied previously, we can define three types of basic priority applying to a business activity:

– To develop *sales opportunities* for the company's current products.
– To improve the *company's competitive* position, in terms of costs and differentiation.
– To apply the *company's resources*, outside its normal activity, by transferring 'assets' in the widest sense.

The four forms of internationalisation can help in satisfying these priorities:

– *Exporting* consists in finding new sales opportunities.
– *Direct foreign investment* results either in increasing sales opportunities (in the case of market strategy), or in improving the company's competitive position (production strategy).
– *Quasi-internalisation* leads either to increasing sales opportunities (market strategy) or to improving the company's competitive position (production strategy).
– *Pure licensing* is a way of using the resources of a company, usually outside its traditional activity.

2. *The Target Markets* The variables which have to be analysed in order to evaluate the target market will be presented under three headings:

1. The *commercial* characteristics will be identified through four indicators:

– *Accessibility to the market* – i.e., barriers in respect of cost price or of quality related to transport.
– *Size of the market*, in relation to the minimum scale efficiency of a production unit (economies of scale).
– *Market development profile* in terms of rate of increase of sales, which corresponds roughly to the position in the product's life cycle.
– *Degree of 'peculiarity' of economic behaviour* (commercial, financial, social, etc.) compared to the company's usual environment.

2. The *industrial* characteristics will be described with four indicators:

– *Local competition*, or more precisely the intensity of competition (in the sense used by M. Porter) in the target market.
– *Technical level* of the main local manufacturers compared with the company's own level.

- *Cost of labour* – i.e., the actual labour cost (salaries plus social contributions plus fringe benefits, per unit produced) compared with international standards.
- *Cost of capital* – i.e., the relative level of interest rates on capital borrowed for local investment.

3. The *political* characteristics will be analysed on four different levels:

- *Trade policy of the recipient state* in terms of entry barriers set up by the government as part of its customs policy (taxes, quotas, regulations, etc.).
- *Policy towards foreign investors*, which can take different forms (tax relief, subsidies, etc.) and which will be judged simply on whether or not it is favourable to direct foreign investment in comparison with other forms of internationalisation.
- *Foreign exchange controls*, considered only from the angle of whether it is possible to transfer revenues back to the home country.
- Lastly, the *political risk* must be evaluated in terms of the probability of the state changing its attitude as regards assets owned by foreigners, and engagements (contracts) signed with them.

3. The Resources Available   There are four categories of resources available to a company for reaching its target markets (marketing, technical, human, financial) which we will examine to find out which form of internationalisation they are most suitable to, considering the target market.

1. *Marketing resources* cover three elements in our model:

- *Product adaptation* – i.e., the fact that the products require no modification in their shape or content to meet the target market's demand.
- *Command of the economic environment* by the company's marketing men, or their capacity to adapt to it.
- *Brand awareness in the target market*, which is a resource which will influence the choice of the form of internationalisation.

2. *Technical resources* will be studied taking six elements into account:

- The *production capacity* available in the company's present plants.
- The *efficiency of the process being used* in terms of productivity, and therefore of costs.
- The *complexity of the company's technical process*, in other words the difficulty of transferring its technology to another company.

- The *age of the technology* – i.e., the time elapsed since it attained a stable form.
- The *pace of technological development* in the industry, which may be relatively slow or fast compared with other industrial sectors.
- The *capacity to adapt the product to local conditions*, which the company has acquired from its experience.

3. *Human resources* will be studied from three angles:

- The *cost of labour* in the home country's production plants compared with those of the target market.
- The *availability of management resources* – i.e., their capacity to devote time to the management of international operations without any repercussions on the company's domestic activities.
- The capacity to *allocate skilled personnel* (technicians, engineers) whose contribution would be needed for the setting-up of the technology in the target market.

4. *Financial resources* will be analysed from two angles:

- The company's *financial capacity* to allocate resources to internationalisation.
- The management's attitude towards *risk taking*, which may of course be more or less conservative.

The Interplay of the Determining Factors

It is now necessary to present the relationship between these determining factors and the forms of internationalisation. We will do this by indicating the characteristics of the target markets and of the resources available which correspond to each of the options of internationalisation.

In actual fact, for the majority of these relationships, we will merely *recall or develop* what has already been pointed out by the other authors or by ourselves. In order to give weight to this model, we will indicate where possible the written source where the relationship under consideration has been studied, totally or in part.

For the sake of clarity and to save time, it seemed to us appropriate to present this information in tabular form. Thus in Tables 9.1 and 9.2, the forms of internationalisation appear vertically, and the determining factors horizontally. The latter will *favour* one or other of the forms of internationalisation, according to their situation. For example, the market size factor will tend to encourage an export strategy when the market is small, a direct foreign investment market strategy when it is large, and a market strategy of quasi-internalisation if it is between the two. These situations must obviously be considered in a relative manner and the table should be read horizontally.

Table 9.1 presents the relationships between the characteristics of target markets and the forms of internationalisation, and Table 9.2 those between the latter and the company's resources. In both cases, the absence of details on the situation of a factor means that it is indifferent to the option under consideration.

The process consists in:

1. Determining the *current priority* among the three which have been selected.
2. Identifying the *main target markets* and studying their characteristics in relation to the activity under study, in order to know whether they call more particularly for one of the four options that have been identified.
3. Lastly, *analysing the company's resources* in relation to each target market, and verifying whether they lead to the same form of internationalisation.

In any event, the choice will have to be based on the form which comes up most often in 2 and 3, as it will be exceptional for all the conditions required by one form to be present together. This choice being clarified, it is now possible to go into technology pricing policies.

9.2   TECHNOLOGY PRICING POLICIES: THE LICENSOR'S POINT OF VIEW

We will now restrict the field of our analysis to the last two forms of internationalisation, quasi-internalisation and pure licensing. We know, from the foregoing model, the motives underlying the strategy which is chosen. This will provide us with a useful base for the *definition of the price of the technology.*

This second part will be devoted to presenting a second prescriptive model relating to pricing policy. This model actually comprises three approaches according to the strategy being chosen: quasi-internalisation through market strategy, or through production strategy or pure licensing. The pure licensing approach being clearly different from the other two, we will consider them separately.

Table 9.1 Characteristics of target markets favourable to each form of internationalisation

		Direct foreign investment		Quasi-internalisation			
Factor	Export	Market strategy	Production strategy	Market strategy	Production strategy	Pure licensing	Sources
Commercial characteristics							
– Accessibility to the market	Easy	Difficult	—	Difficult	—	—	Aitelhadj–Bidault (1980)
– Size of market	Small	Large	—	Medium	—	Medium	Aliber (1970) Vernon (1966)
– Market development	Emergence	Growth	—	Maturity	—	Decline	Telesio (1979)
– Degree of 'peculiarity'	Low	Low	—	High	—	High	Ragazzi (1973)
Industrial characteristics							
– Local competition	Weak	Medium	Strong	Strong	Strong	Strong	Telesio (1979)
– Technological level	Medium	—	—	Low	Low	Medium	Aitelhadj–Bidault (1980)
– Cost of labour	High	Medium	Low	Medium	Low	—	Michalet (1976)
– Cost of capital	High	Low	Low	Medium	Medium	—	
Political characteristics							
– Trade policy	Liberal	Protectionist	Promotion of exports	Protectionist	Promotion of exports	—	Aitelhadj–Bidault (1980)
– Policy towards foreign investors	—	Favourable	Favourable	Unfavourable	Unfavourable	Unfavourable	Emmanuel (1981)
– Foreign exchange policy	Not very liberal	Very liberal	Liberal	Very liberal	Liberal	Liberal	Buckley and Casson (1976)
– Political risk	Medium	Low	Low	High	High	High	Emmanuel (1981)

Table 9.2 Characteristics of available resources favourable to each form of internationalisation

| | | Direct foreign investment | | Quasi-internalisation | | | |
	Export	Market strategy	Production strategy	Market strategy	Production strategy	Pure licensing	Sources
Marketing resources							
– Adaptation of product	Strong	Medium	Weak	Medium	Weak	Medium	Ragazzi (1900)
– Command of environment	Strong	Very strong	Medium	Weak	Weak	Weak	
– Local brand awareness	Strong	Strong	Weak	Medium	Weak	Medium	Caves (1972)
Technical resources							
– Available production capacity	High	Medium	Low	Medium	Low	–	Buckley and Casson (1976)
– Efficiency of process	High	High	High	Medium	Medium	High	
– Easiness to imitate	–	High	High	Medium	Medium	High	Mason (1973)
– Age of technology	Low	Medium	High	High	High	Low	Magee (1978)
– Pace of development	Medium	Medium	Slow	Slow	Slow	Fast	Teece (1981)
– Capacity to adapt	Weak	Strong	Strong	Medium	Medium	Medium	
Human resources							
– Cost of labour	Low	Medium	High	Medium	High	–	Michalet (1976)
– Management availability	Low	High	High	Medium	Medium	Low	Cremadez (1980)
– Capacity to allocate personnel	Low	High	High	High	High	Medium	Cremadez (1980)
Financial resources							
– Financial capacity	Low	High	High	Low	Low	Low	Marois (1980) Telesio (1979)
– Attitude towards risk	Acceptance	Acceptance	Acceptance	Refusal	Refusal	Refusal	Emmanuel (1981)

A Pricing Model for Pure Licensing Strategies

This first case in fact concerns the strategy which has attracted most attention on the part of research scholars and experts. Hence the principle of price determination is the best known, even though it continues to be the subject of debate.

Our objective here is merely to present a synthesis of the main contributions to this pricing policy (see Figure 9.2). It should be pointed out that in doing this, we leave a certain number of questions unanswered, like those relating to the 'coefficients'. What coefficient should be applied, for example, to determine the share which should be claimed from the licensee's profit? Similarly, what coefficient should be used to determine technology payments as a multiple of transfer costs? We have to recognise that today no sound, convincing arguments exist to answer these questions. Let us hope that in the future research will be devoted to these questions.

Nonetheless, the suggested procedure can be regarded as a method which is more relevant than the usual reference to the industry standard royalty rates. Even if the *LSEP* and the multiple are not known in detail, it is still possible to work with intervals, which at least provides a framework for the negotiation.

A Pricing Model for Quasi-internalisation Strategies

The model we are putting forward here is not based solely on the objective observation of facts. It expresses preferences, some of which refer to what is most commonly practised, and therefore the most 'normal'. But it also embodies a few options which are contrary to the behaviour of most firms. So we will have to explain these preferences as we proceed.

The first principle is common to the two last parts of the model, one applying to the market strategy, the other to the production strategy. This principle consists in making the technology's *price*, and the licensing agreement's *motives*, coherent. In practice, it means fixing this price in order to cover at least the transfer cost, which really comes to drawing practically no net return from the technology transfer. This means that most of the remuneration results from profits made on commercial transactions (purchases from, and sales to, the licensee) while transfer costs are covered in case the cooperation does not materialise.

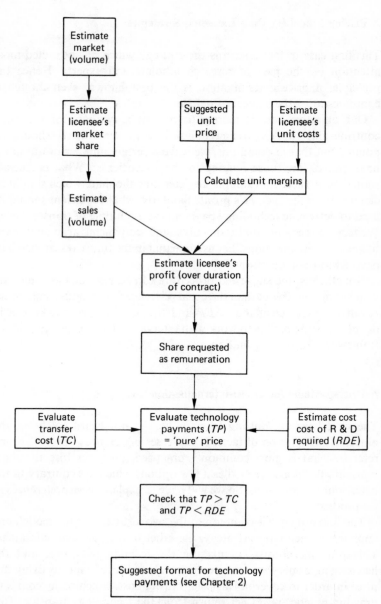

Figure 9.2   The steps in the determination of the price of a technology, in the case of a pure licensing strategy, from the licensor's point of view

The licensor then sells his technology at a price which is lower than what he would charge if he were following a pure licensing strategy (Figure 9.2). So he has to make sure that what he loses by lowering the price is at least compensated for by the profit he makes on sales.

It could be objected that this form of remuneration can create distrust in the relationship, insofar as it is difficult for the licensor to check what he is paying. For this reason, we believe that the licensor, if he wants to build up a lasting (and therefore efficient) relationship, should be able to provide figures justifying a *'normal' remuneration* in the market under consideration. This could be done by calculating the contribution to the total margin of the licensor of the various components of the finished product. A corrective coefficient could then be applied to this 'normal' margin, to take into account the specific marketing conditions (in the form of intermediate products). In fact calling in an impartial expert to confirm these evaluations could be a guarantee to the licensee and a selling argument for the licensor.

Having made that clear, we can now describe the essential steps in the determination of price for a market strategy, then for a production strategy.

The Determination of Price with a Market Strategy

The approach consists essentially in evaluating the profit margins which can be obtained on sales to the licensee during the whole duration of the contract and to compare the discounted figures, on the one hand, with the 'pure' price of the technology (the price he would have fixed had he followed a technology strategy), and on the other with the transfer cost (see Figure 9.3).

The licensor must indeed make sure that he is earning by this method *at least as much* as if he had merely sold his technology. In addition, he must allow himself a margin on the transfer cost, because in spite of profit margins being roughly the same value as the expenses incurred to transfer the technology, he runs the risk of not covering them in case the contract is broken. So the final price should, in any event, be at least equal to the transfer cost, and all the higher as the expected commercial margins are low. In conclusion, the licensor must make sure that the sum of revenues (profit margins plus technology payments) is at least equivalent to the 'pure' price, which is of course a rather approximate variable, as we saw in our analysis of pricing policy in a technology strategy.

174

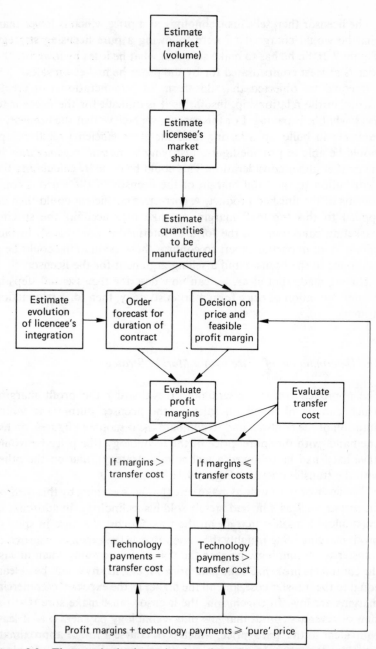

Figure 9.3   The steps in the determination of a technology's price: the case of a market strategy, from the licensor's point of view

The Determination of Price With a Production Strategy

In a totally symmetrical manner, we must find out here if the savings brought about by the licensee's involvement are of sufficient size to justify the operation (see Figure 9.4).

Whatever policy is adopted, it should be clearly understood that we cannot totally eliminate the irrational element in the determination of a technology's price – i.e., it is not a microeconomic act, but the result of the action of a large number of agents. Without claiming that prices are fixed according to the 'laws of the market', it should not be forgotten that a form of competition does exist. This competition can show peculiar characteristics owing to the very nature of the exchange itself: technology. The various suppliers can follow diverging strategies, as we have seen, and the result of the competitive clash will then produce a price which is consistent with none of the strategies being followed.

The licensor's behaviour having been analysed in a normative manner, it is time to consider the other point of view, that of the licensee.

9.3 TECHNOLOGY DEVELOPMENT STRATEGIES AND PRICING POLICIES: THE LICENSEE'S POINT OF VIEW

We now have to draw the implication for the licensee's strategy of what we know of the licensor's pricing policy. To this end, we must return to the strategic dimension underlying the agreement, by analysing how each partner's desire to co-operate comes to meet that of his opposite number. One important point for the licensee will be to determine the licensor's motives, and, if he accepts them, to deduce an appropriate pricing strategy.

Technology Development Strategy and Type of Licence

Our argument thus consists in pointing out that the strategy for price negotiations must be defined taking into account the type of licensing agreement being entered into.

While we can expect the acquisition of technologies to be consistent with the priorities chosen by the potential licensee, the objectives he is aiming at (technology, market and production) are not sufficient to determine a pricing policy. One has to know the framework of the

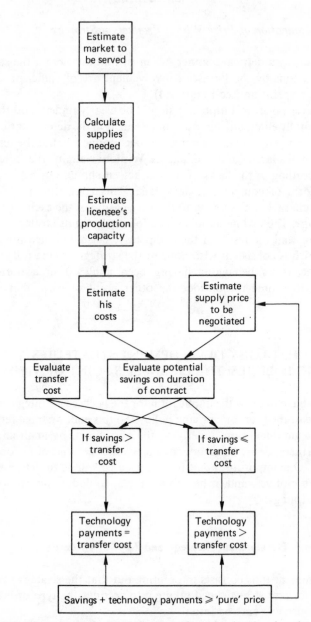

Figure 9.4   The steps in the determination of a technology's price: the case of a production strategy, from the licensor's point of view

licensing agreement resulting from the interaction between the licensor and the licensee, which means that the generic strategy adopted by the partner must be taken into account.

We know that the licensee may wish to 'purchase' a technology for a variety of reasons (see Chapter 6) such as, for instance, the improvement of his productivity on products he already manufactures, entry into a new market for reasons of diversification, or additional contributions to his research and development results. But these motives can give rise to widely different licensing agreements.

Thus the objective of entering an unknown market with a new technology can lead to a 'pure' licence when the subject of the exchange is merely information and knowledge. As opposed to that, such an objective can also result in entering into a relationship with the licensor for the supply of equipment and intermediate products, which is the case with agreements of the 'licence + purchase' type. Lastly, if the market in question is precisely that of the licensor, the licensee may wish to use his partner's logistics to support the marketing of his products. In other words, the type of technology transfer a licensee will be prepared to sign has to be determined over and above the motivation of his objective. The type of agreement will thus eventually depend on the match between the resources sought by the licensee on the one hand, and the strategic motives of the licensor on the other (see Chapter 7).

It is only after this analysis has been carried out that a policy for price negotiation can be formulated. The model we propose offers three facets, each corresponding to a type of agreement (see Figure 9.5).

Technology Pricing Policies for Licensees

We will present the three elements of the pricing model from the licensee's angle, starting with licensing agreements with commercial transactions attached, before going on to the case of a 'pure licence'.

A Pricing Policy for a 'Licence + Purchase' Agreement

The criterion for decision in such a case is always, after all, *the return on investment ratio* – i.e., the ratio of the discounted cashflow on the investment brought about by the licensing agreement. The agreement will be signed if this ratio is greater than, or equal to, the standard which the licensee has defined. But there can be many interferences, as the licensor in fact draws on the cash-flow at different levels. First through technology payments, but also through supply prices (see Figure 9.6).

So these two variables are at the heart of the negotiation. In a manner which is symmetrical to the model recommended for the licensor, and from a different angle, the licensee should strive to estimate transfer costs on one hand, and supply costs on the other, possibly with the help of an expert accepted as such by the two parties.

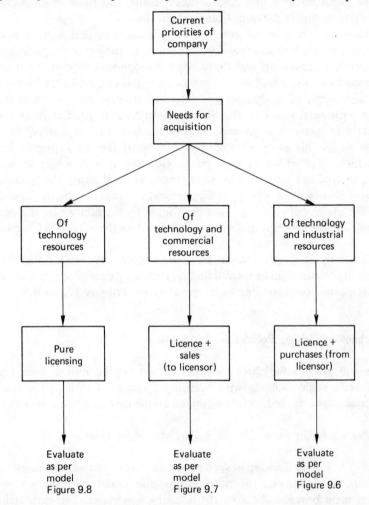

Figure 9.5 The relationship between the strategy for the acquisition of a technology and the evaluation of technology payments

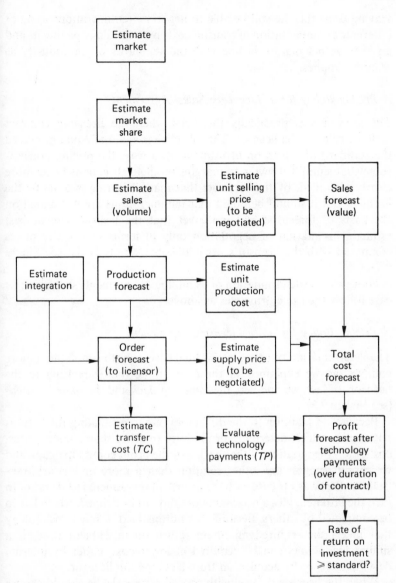

Figure 9.6 A model for evaluating the price of a technology, from the licensee's point of view, in the case of a 'license + purchase' agreement

Having done this, he will be able to negotiate remunerations using as criteria the compensation of transfer costs by technology payments and an average net margin in line with the standards of the industry as regards supplies.

A Pricing Policy for a 'Licence + Sales' Agreement

The approach is practically the same when the licensing contract includes sales to the licensor. The objective is again to equal or exceed the standard of return on investment. In a way, the pricing policy is somewhat simplified owing to margins on the transactions being more clearly identified, by the very fact that manufacturing is done by the licensee. But it should be noted that remunerations are negotiated on the basis of forecasts, and moreover that the licensor can at best estimate his margin of negotiation only in terms of transfer prices compared with the licensor's other possible supply costs (see Figure 9.7).

However, the determination of technological payments should in this case follow the same principles as above.

A Pricing Policy for 'Pure Licensing' Agreement

The determination of price, in this final case, is both simpler and more compex, as we explained in the case of the model relating to the licensor. In fact, we find here the *same approach* and the *same questions* (see Figure 9.8).

Because, in addition to the difficulties met in estimating the various parameters which determine profitabilty (market share, sales, prices, costs, etc.) the question of the sharing of the profit comes up again, and we cannot suggest any other solution than a more or less arbitrary 'standard' as was the case with the return on investment (ROI) ratio. In fact, the latter enables a necessary condition to be defined, which has to be completed by others. Because it is not justified to buy a technology that meets the required minimum return on investment ratio, if a similar technology can be acquired at lower cost, either by internal development, or by acquisition from a competing licensor.

When the licensee is faced with several proposals, he should choose to acquire the technology:

– Among those offering a *sufficient return on investment ratio*.
– That which entails the *lowest share of profit being paid to the licensor*, if several offer comparable profitability (after technology payments).

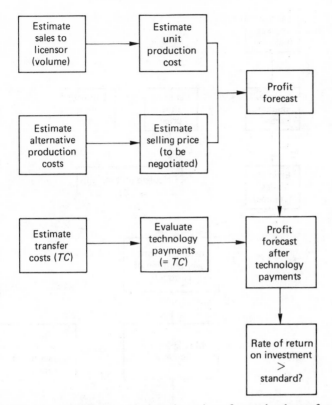

Figure 9.7 A model for evaluating the price of a technology, from the licensee's point of view, in the case of a 'license + sales' agreement

But above all, it should be borne in mind that in such competitive situations the strategies being followed by the potential licensors can be divergent, which will effectively cause differences in the variables mentioned. The licensee can obviously use such differences to strengthen his bargaining power. However, we do not believe that he should be short-sighted by automatically selecting the 'cheapest' proposal, with the risk of entering into a relationship having a configuration very different to what he was looking for. On the contrary, he should stick, as we explained when dealing with the technology development policy, to this type of technological transfer which corresponds best with his strategy and especially with the resources which he needs.

182

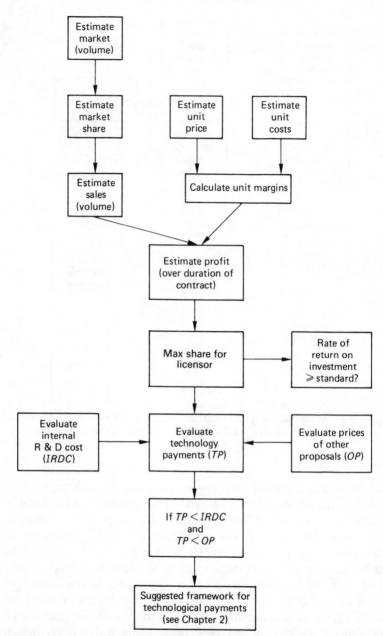

Figure 9.8 A model for evaluating the price of a technology, from the licensee's point of view, in the case of a pure licensing agreement

Conclusion

Technical knowledge, defined as a body of data and rights necessary for the production of goods, is the subject of transactions. The supplier firm (or licensor) and the recipient firm (or licensee) may or may not be in the same industry – i.e., companies carrying out the same activity on different markets. We became interested in this type of transaction in order to study how prices for the transfer of technical knowledge are determined. In order to do this, we have had to concentrate exclusively on transfers between firms which are financially independent from each other, considering the bias which may intervene in the determination of prices between parent companies and their subsidiaries.

We have attempted to identify, under these conditions, the *determining factors in the fixing of technology prices*, founding our work on theoretical reflection, and on our empirical analysis of economic data drawn from actual contracts.

Measuring the price of a technology raises considerable difficulties. While a theoretical definition of price can indeed be found without great difficulty, the statistical measurement comes up against obstacles due to companies' behaviour in the field of remuneration. In a transfer of technology, there are many forms of remuneration. Only a few concern the technological knowledge itself, the others relate to the ancillary services which precede or accompany the setting up of the technology in the recipient firm. But this theoretical distinction is not often applied in practice, as the negotiators carry out a kind of implicit compensation between the different forms of remuneration. Thus the price should be considered as the sum of payments made by virtue of the licensing contract, after deducting the cost of services provided for the implementation of the agreement. This price is of course an absolute value and therefore cannot be used in comparing different contracts. Two 'ratios' can be used to do this: the first one, called the 'multiple', establishes the relationship between technology payments and transfer costs, while the other, called '*LSEP*' evaluates the share of the licensor in the profits made by the licensee.

Traditional theory, inspired by the neoclassical school of microeconomics, leads one to believe that there could be a single price for each technology. Of course this is not the case, not only owing to the uncertainty underlying this type of transaction, but particularly because the behaviour of firms does not always follow the hypotheses

183

which have been laid down. The level of the price which will effectively be paid depends partly on the strategy being followed by the licensor, and on the content of the agreement.

Now this point has not been fully appreciated by people like Contractor and the UNIDO team, who have worked on the question of technology pricing. They have, in a sense, tried to define the determining factors, while disregarding differences in strategy between licensors. Although their results are most interesting, and in part complementary, we have considered that they were addressing only part of the problem.

They consider implicitly in effect that the licensor's essential motivation is to draw profits from the sale of his technology. We show that this motivation applies only to one *specific case* – i.e., where the supplier firm is following what we call a technology strategy. On the contrary, where the strategy being followed is market- or production-oriented, the pricing policy is completely different, and the licensor is seeking something beyond a good remuneration. In the case of a market strategy, his aim is to find new outlets. In the case of production strategy, he is trying to improve his supplies. He makes use of the transfer of his technologies to attain his objectives more easily.

In a licensing relationship, the licensor is obviously not alone. We therefore have to show how his motives can be compatible with those of the licensee. Hence we show that the latter's expectations often go beyond the mere supply of technologies, and sometimes extend to resources of a material, commercial or financial nature. It follows that the transfer of a technology does not always result in the plain transmission of knowledge to an outside agent, but that the licensor can go on controlling an essential part of his 'technology asset'. We then make the distinction between *'pure licensing'*, where the exchange between the two partners is limited to technological information and its ancillary rights, and *'quasi-internalisation'*, where the transfer of technologies is not complete, but gives rise to a flow of complementary transactions.

Consequently, we have developed the hypothesis that these differences in strategy must have an effect on pricing policy: licensors negotiate their remuneration according to their objectives and do not systematically attempt to get the optimum price for their technologies.

Pricing policy is first analysed *at a theoretical level*, and we consider that the licensor can derive more profit by increasing the price of intermediate products to the licensee, even if this means reducing the amount of royalties to be drawn. We even think that a zero-price policy is possible, if sufficient margins are obtained on the ancillary transactions. We thus envisage policies which differ from those which are traditionally accepted. But that meant checking that such policies were indeed being followed.

So we have attempted to validate our hypothesis by analysing two samples: one made up of 33 licensors, the other of 29 licensees. Considerable differences appeared between contracts, according to the type of strategy being followed by the licensor. To summarise, it can be said that the ratio of technology payments to transfer costs (the multiple) is lower when the licensor has market or production motivations. In this case, the ancillary transactions are a determinant factor of price, and transfer costs no longer play such an exclusive part.

Lastly, we have presented the conclusions obtained for the foregoing analyses in the form of a *model for choosing the type of international development, and the pricing policy of technologies*. The first part of the model provides a list of factors which are favourable to each one of the four forms of internationalisation which we have identified: export, direct foreign investment, pure licensing and quasi-internalisation. For each one of them we recommend a method for calculating prices which takes the strategic motivations into account. Each method is presented from the two points of view: that of the licensor, and that of the licensee.

These results concern essentially the negotiators of the two parties to a licensing contract. However, they may also enlighten the public agencies which control technology transfers in most countries. Indeed the political implications of our approach are far from negligible. They express the need not to consider transfers of technologies as something homogeneous, and to take into account the differences which lie therein. This pleads in favour of an individual treatment of these operations, which does not rely on rules which are too general, and which takes the *partners' motives* into account, with their corresponding pricing policies. We have seen that the recipient country's economic and technical environment has an influence on the type of licensing agreement, and favours one type of generic strategy or another. Let us remember especially that market and production strategies are more common in transfers to developing countries than to industrially advanced nations. But it needs to be emphasised that none of the three strategies is in itself preferable. Each has advantages and disadvantages. The balance between these two dimensions depends mainly on the situation of the country's and the potential licensee's resources, but also on the priorities which the state and the firm have fixed. We can recommend only a *detailed investigation* of the business, involving a clarification of the positive and negative points in each type of licence, as seen by the public authorities.

But we must recognise that there is a long way to go before this new approach to the licensing relationship can be accepted. Our work comes up against a *certain number of limits* which we would like to mention. First of all, our working hypotheses have been validated on samples of somewhat limited size, even if they follow or exceed the standards in this field. We have also limited our investigation to a small number of countries, and the extension of our conclusions to other countries still remains to be done. On the other hand, it is regrettable that some essential data was not available. Notably, it has not been possible to make a strict comparison of the shares of the licensors in the profits of the licensees (the *LSEP* ratio).

Yet this ratio expresses a most interesting concept in technology pricing, but because we did not have sufficient responses on this point, we were unable to calculate it for most of the contracts. Consequently we could not really see if market and production strategies showed a lower ratio than technology strategies. This is all the more disappointing because the small volume of data which was available seemed to give some evidence on this point.

Lastly, it should be borne in mind that our research was devoted essentially to horizontal and external 'transfers'. We deliberately ignored the *other forms of transfer*, mainly for methodological reasons. Yet is the validity of our results restricted only to those operations which we selected? We have not treated this question directly. So all this opens up new *research opportunities*.

The first one would be to apply the same method to other samples of licensors and licensees in other countries. However, we do not believe that by doing this, a sufficient number of cases could be assembled in one survey to allow a real validation. Another way would be to analyse the data gathered by public organisations controlling technology transfer, which request the licensee to provide detailed economic forecasts before giving their approval.

Yet another approach would consist in examining the pricing policies followed for the *other types of technology transfer*. This would allow a study of the possible differences with the horizontal and external transfers. Vertical transfers between Research and Development and manufacturing are not well known from this angle. Most of them certainly take place internally, but a few private and public research centres sell technologies which they have developed. In the same way, progress must be made concerning the pricing policy of engineering consultants who act as coordinators and hold a specific place in the international flow of technologies. It would also be interesting to measure more precisely the discrepancies between prices charged in internal transfers, and those applied between independent partners. The policy of multinational firms is indeed at the heart of a debate on overpricing in internal transfers of goods between their plants. In the absence of a theory of technology pricing, there exists no reference to make any sort of statement on the way in which multi-nationals invoice their technologies internally. The managers of these companies themselves have very few management rules at their dis-posal to help them make their decisions. It is to be hoped that research will be carried out to adapt the models of technology pricing to cases of internal transfer. Lastly, a related field deserves more attention than it has received on the question of pricing: franchising. This form of commercial development is going through a period of exceptional expansion, particularly in the distribution sector, but the principles of remuneration, probably close to those of manufacturing technologies, have not been the subject of much study.

Considering the increasingly important position of information in industrial systems, the *commercial exchange of information* should play a major part in the coming years. Whether it be technical or other types of information (marketing, legal, fiscal, etc.), all companies are acquir-ing more and more information. Of course, most of it is public. But in the increasing number of contracts between firms for licences or franchises, between firms and universities or technical centres, between firms and consultants, etc. the information being transmitted is private in nature. It seems that the trends in this area are more distinct than those relating to the international flow of manufacturing technologies. Hence it is proving necessary to understand the working of these extended exchanges of information, which more and more firms are getting involved in. At the end of our reflection, we remain convinced that purely rational approaches will have to be discarded, and that the interplay between market forces and the dynamics of organisation will have to be taken into account.

Notes and References

Introduction

1. UNIDO Secretariat (1983).
2. OECD (1981) p. 99.
3. Perlmutter and Sagafi-Nejad (1981) p. 10.
4. Vaitsos (1974).
5. Contractor and Sagafi-Nejad (1981).
6. Only English-language publications are listed, although documentary sources, especially in French and Spanish, are considerable. See Joly (1981).
7. Madeuf (1981) pp. 173ff; Caves and Uekusa (1976) pp. 125ff; Ozawa (1981).
8. This viewpoint was expressed recently in a publication of the joint ESCAP/UNCTC research unit, under the title 'Costs and Conditions of Technology Transfer through Transnational Corporations' – see the summary which appeared in the *TIES Newsletter*, 26 (UNIDO, July 1984) p. 9.
9. Baranson (1970) p. 470.
10. Germidis (1977).
11. Mason (1973).
12. Teece (1981).
13. Mason (1973) p. 8.
14. Baasche and Duerr (1975) Chapter 2.
15. Perlmutter and Sagafi-Nejad (1981) Chapter 7.
16. Remark already made by Lovell in 1969 – see Lovell (1969).

1 International Technology Transfer: Substance and Framework

1. Gonot (1974), quoted by Madeuf (1981) p. 69.
2. Hall and Johnson (1970).
3. Arrow (1962), used in Lamberton (ed.) (1971).
4. Magee (1978).
5. Boulding (1971), used in Lamberton (ed.) (1971).
6. Magee (1978) p. 321.
7. Contractor (1981a) pp. 13–14; Orleans (1981) pp. 320–4.
8. OECD (1981) p. 20.
9. Boulding (1971) p. 23.
10. Radner (1979) p. 4–3.
11. Radner (1979) p. 4–3.
12. Arrow (1962) p. 147.
13. OECD (1981) p. 20; Emmanuel (1981) p. 73; Madeuf (1981).

14. Radner (1979) p. 2–2; Arrow (1970); Teece (1981) p. 85.
15. Arrow (1970) p. 275
16. See especially: Weil (1980) and Magee (1978) p. 333.
17. Dunning (1982) p. 10.
18. See in particular the segmentation suggested by Marois (1980) p. 12.
19. See for example: Teece (1976).
20. Aitelhàdj and Bidault (1983) pp. 83–91.
21. Carlier (1977).
22. Teece (1976) p. 23.
23. Michalet (1976) pp. 189ff.
24. Madeuf (1985).
25. Janisewski and Besso (1982) Chapter 1, p. 7.
26. Vaitsos (1974) Chapter 6.
27. Contractor (1981a) Chapter 5; Kopits (1976), quoted by Contractor (1981); Telesio (1979) p. 3.
28. Teece (1976) p. 18.
29. Blanc (1980).
30. Madeuf (1985) pp. 85ff.
31. Buhler, Carrière, Petit and Sitbon (1981); Lovell (1969); Contractor (1981a).
32. Gaudin (1982) p. 20.
33. Contractor (1981a) p. 146; Lovell (1969) Chapter 2; Buhler *et al.* (1981).
34. See in particular that done by Bernadette Madeuf in France.
35. Janisewski and Besso (1982) p. 6.
36. Perlmutter and Sagafi-Nejad (1981) p. 207.
37. Madeuf (1985) p. 84.
38. Madeuf (1985) p. 9.
39. Huard (1974) Annexe A, pp. 177–224; Chevalier (1977) pp. 79–104.
40. Steedman (1984).
41. Bizec (1981) p. 43.
42. Lovell (1968) p. 27; Weil (1980); UNIDO (1983).
43. The size of our samples is a good average for this type of research, where it is usual to work on 30 to 100 cases: David J. Teece (26 cases), Piero Telesio (66 companies), Farok Contractor (39 companies with 102 contracts).

2 Types of Remuneration and Forms of Payment in International Technology Transfers

1. Janisewski and Besso (1982) pp. 3–63.
2. See also Modiano (1983) pp. 553–91, which is based on the *Licensing Law Handbook*.
3. Perlitz (1980) pp. 75–82.
4. Bowler (1980) pp. 241–7.
5. Vaitsos (1974), especially Chapter 1.

3 The Pure Theory of Technology Pricing

1. The following examples can be quoted: Bowler (1980) pp. 241–7; Orleans (1981) pp. 320–4.
2. Quoted by Gaudin (1982) p. 51.
3. Scherer (1980) p. 441.
4. Scherer (1980) pp. 444ff.
5. Johnson (1970) p. 37.
6. See the paragraph in this chapter where we deal with transfer costs.
7. See Penrose (1973) pp. 768–85 and Lall (1976) pp. 1–15. Penrose and Lall come to the conclusion that less-developed countries should challenge not so much the patent system as the existence of non-competitive structures which allow restrictive practices.
8. Piatti (1985) p. 116.
9. Orleans (1981) p. 320.
10. Caves and Murphy (1976) pp. 572–86.
11. Vaitsos (1974) p. 140.
12. The company's profit is the difference between total costs (CT) and total revenue (RT), itself the product of quantity sold (Q) and price (P). We assume that a relation exists between cost and quantity on the one hand, and price and quantity on the other.

    We have:

    $$CT = f_1(Q) \text{ and } P = f_2(Q)$$

    So we have:

    $$\pi = RT - CT, \text{ where } RT = Q \cdot f_2(Q) = f_3(Q)$$
    $$\pi = f_3(Q) - f_1(Q) = f_4(Q)$$

    Max. π requires $f_4'(Q) = 0$ and $f_4''(Q) < 0$
    The first condition is expressed by $f_3'(Q) - f_1'(Q) = 0$
    Hence $f_3'(Q) = f_1'(Q)$
    $f_1'(Q)$ is the increase in total cost arising from an infinitesimal increase in Q. It is called marginal cost.
    $f_3'(Q)$ is the increase in total revenue arising from an infinitesimal increase in Q. It is called marginal revenue.
13. Caves and Murphy (1976) p. 577.
14. Bowler (1980) p. 244.
15. For a critical presentation of this theory, see Angelier, Benzoni, Bidault *et al.* (1983) Chapter 7.
16. Mahieux (1981) pp. 65–75; Graham (1982); Magee (1978) pp. 319–40.
17. Magee (1977) pp. 203–24.
18. See, among others, on this question Morvan (1976).
19. So many opinions have been expressed on this subject that it would be impossible to quote all of them. See, among others: Lovell (1969) p. 34; Teece (1976) pp. 86–8.
20. Orleans (1981) p. 320.

21. Finnegan and Mintz (1978) pp. 13–23.
22. Vaucher (1978).
23. Mougeot (1979). A concise presentation is also made by Bidault and Zervudachi (1983).
24. Finnegan and Mintz (1978) p. 19.
25. Vaucher (1978).
26. Lovell (1969) p. 24.
27. Riveline (1983) p. 66.
28. INTI (1983).
29. Arrow (1962), used in Lamberton (ed.) (1971) p. 147; Silberston (1967), used in Lamberton (ed.) (1971) p. 225.
30. Teece (1976).
31. Teece (1976) p. 36.
32. Teece (1976) p. 44.
33. The results of statistical tests on these hypotheses are presented in Teece (1976) Chapter 4, pp. 57ff.
34. Finnegan and Mintz (1978) p. 18.
35. Contractor (1981a).

4 Two Approaches to Pricing Policy

1. Ozanne and Hunt (1971), quoted by Caves and Murphy (1976) p. 577.
2. We will mention only Finnegan and Mintz (1978) pp. 13–23; Vaucher (1978).
3. A synthetic presentation of the UNIDO method will be found in Bidault and Zervudachi (1983) pp. 53–60.
4. There are several presentations of this model: Janisewski and Besso (1982) pp. 3–67; UNIDO (1982); Arni (1984).
5. We will subsequently abandon this hypothesis.
6. UNIDO (1982) p. 48.
7. UNIDO (1983b).
8. Cf. two versions of their study: Instituto do Investimento Estrangeiro (1983); UNIDO (1983).
9. UNIDO (1983b) p. 12.
10. UNIDO (1983b) p. 11.
11. Contractor (1981a).
12. Root and Contractor (1981) pp. 23–32; Root (1981).
13. Root and Contractor (1981) p. 26.
14. Bidault and Mullor (1984).
15. March and Simon (1964).
16. Cf. James D. Goodnow's commentary on Contractor's work (1981a).
17. See, for example, Conseil et Développement (1980).

5 Licensing Out Strategies

1. Andrews (1980).
2. Aitelhadj and Bidault (1979), (1980) pp. 94–102; (1981b) pp. 37–45; (1982) pp. 39–52.
3. OECD (1981) Chapter 1; UNIDO (1983).
4. See, in particular, Alain Weil (1980) p. 46; Lovell (1969); Contractor (1981b) p. 79.
5. *Les Echos* (11 October 1984).
6. *Le Moniteur du Commerce International*, 228 (3 April 1978).
7. Duran and Drouvot (1978).
8. Aitelhadj and Bidault (1981b).
9. Vaitsos (1974) p. 52.
10. Aitelhadj and Bidault (1981b) p. 43.
11. Telesio (1979) p. 17.
12. Lovell (1968) p. 14.
13. Lovell (1969) p. 50.
14. See Chapter 4.
15. Vaitsos (1974) pp. 45–6.
16. Lovell (1969) p. 51.
17. Conseil et Développement (1980).
18. Aitelhadj and Bidault (1981a).
19. Koerber (1980).
20. For a more complete analysis, see Angelier, Benzoni, Bidault *et al.* (1983) pp. 258ff.
21. *Business Week* (1 October 1984, p. 75).
22. Savary (1981) p. 123.
23. See principally: Telesio (1979) p. 13; UNIDO (1983b).
24. Lovell (1968) Chapter 2.
25. Some of them have even created a specialised subsidiary company – e.g., TECNOVA for Pechiney, DITT for Electricité de France.
26. Aitelhadj and Bidault (1979) p. 14; Aitelhadj, Bidault and Sierra (1982).
27. For the presentation of this list, we classified the items in random order.
28. It must be remembered that the licensees in the second sample are not the partners of the licensors in the first sample.
29. Contractor (1981a) p. 37.
30. Telesio (1979) p. 37.
31. Detrie and Ramanantsoa (1983) p. 16.
32. Lovell (1968) Chapter 2.
33. For an extensive presentation of these debates, see Huard (1974) pp. 177ff.
34. Jacquemin (1974) pp. 145–57.
35. Jacquemin (1974) p. 148.
36. Michalet (1976) pp. 150–4.
37. Dunning (1980) pp. 9–31.

6 Licensing In Strategies

1. Lovell (1968).
2. We should particularly like to thank Mr Bendjillali of the Office de Développement Industriel (ODI) of Rabat, Morocco.
3. Porter (1985) pp. 183ff; Rothschild (1984) p. 111.
4. Rothschild (1984) p. 111.
5. Telesio (1979) p. 14.
6. Caves, Crookell and Killing (1983) pp. 254–7.
7. Caves, Crookell and Killing (1983) p. 255.
8. UNIDO (1983) p. 5; Sagafi-Nejad (1979).
9. See *Business Week* (1 October 1984).
10. Telesio (1979) p. 61.
11. Mansfield (1983) p. 24.
12. Killing (1975), quoted by Bonin (1985).
13. Se-Jung Yong and Lasserre (1982).
14. Vaitsos (1974) p. 135.
15. Yong Hee Chee (1979) p. 58.
16. Aitelhadj, Bidault and Sierra (1982).
17. Alain Weil (1980).

7 A Theory of the Licensor–Licensee Relationship: the Quasi-internalisation Model

1. Coase (1937).
2. Williamson (1975).
3. Alchian and Demsetz (1972) pp. 777–95.
4. Buckley and Casson (1976).
5. Teece (1981) pp. 81–96.
6. Caves, Crookell and Killing (1983) pp. 249–67.
7. Bonin (1985).
8. Coase (1937) p. 5.
9. Williamson (1975) p. 118.
10. Buckley and Casson (1976) p. 41.
11. Houssiaux (1957) pp. 385–411.
12. Porter (1985) p. 57.
13. Caves and Murphy (1976) pp. 572–86.
14. Perroux (1973).
15. Jacquemin (1967), (1974) pp. 145–57.
16. Vaitsos (1974).
17. Vaitsos (1974) p. 47.
18. Caves, Crookell and Killing (1983) p. 250.
19. Coase (1937) p. 4.
20. Buckley and Casson (1976) pp. 36–45.
21. Teece (1981) pp. 82–4.
22. Teece (1981) p. 85.
23. Buckley and Casson (1976) p. 39.

24. Buckley and Casson (1976) p. 43.
25. Teece (1976).
26. Teece (1976) p. 91.
27. Buckley and Casson (1976) pp. 41ff.
28. Caves and Murphy (1976).
29. Buckley and Casson (1976) p. 44.
30. Caves and Murphy (1976) p. 575.

Part III   Introduction

1. Conseil et Développement (1980).
2. Lovell (1969) p. 27.
3. Bidault (1986).

8   An Empirical Analysis of Technology Pricing

1. Vaitsos (1974); Contractor (1981) p. 41.
2. Contractor (1981a) p. 34.
3. Contractor (1981a) p. 37.
4. Contractor (1981a) p. 116; cf. equations 5.1 and 5.2 on one hand and 5.6 and 5.7 on the other.
5. This is the operating profit realised on commercial transactions with the licensee, and computed by those interviewed on an annual basis, and then over the whole duration of the contract.
6. ISEA by John Eaton on HP3000.
7. The *t*-value obtained is considerably less than 2.

Bibliography

AITELHADJ, SMAÏL and BIDAULT, FRANCIS (1979) *Les PMI régionales et les échanges technologiques*, Institut de Recherche de l'Entreprise. Lyon-Ecully.

AITELHADJ, SMAÏL and BIDAULT, FRANCIS (1980) *L'insertion des PMI dans la nouvelle Division Internationale du Travail*, *Revue d'Economie Industrielle*, 14 (4) pp. 94–102.

AITELHADJ, SMAÏL and BIDAULT, FRANCIS (1981a) *Catalogue des technologies exportables de la région Rhône-Alpes*, Institut de Recherche de l'Entreprise, Lyon-Ecully.

AITELHADJ, SMAÏL and BIDAULT, FRANCIS (1981b) *Stratégies de transfert de technologie*, *Le Moniteur du Commerce International (MOCI)*, 451 (18 May) pp. 37–45.

AITELHADJ, SMAÏL, BIDAULT, FRANCIS and SIERRA, GILLES (1982) *Transfert de Technologie: 35 PMI s'expliquent*, *Le Moniteur du Commerce International*, 489 (8 February).

AITELHADJ, SMAÏL and BIDAULT, FRANCIS (1983) *Transfert de technologie: les aléas de la concurrence en retour*, *Revue Française de Gestion* (March–April) pp. 83–91.

ALCHIAN, ARMEN A. and DEMSETZ, HAROLD (1972) Production, Information Costs, and Economic Organization', *American Economic Review*, 62, pp. 777–95.

ALIBER, ROBERT Z. (1970) 'A Theory of Direct Foreign Investment', in Kindelberger (ed.) (1970) pp. 17–34.

ANDREWS, KENNETH R. (1980) *The Concept of Corporate Strategy*, Homewood, Ill., Richard D. Irwin (2nd edn).

ANGELIER, JEAN-PIERRE, BENZONI, LAURENT, BIDAULT, FRANCIS *et al.* (1983) *Rente et structure des industries de l'énergie*, Collection Energie et Société, Grenoble, PUG.

ARDISSON, JEAN-MARIE and BIDAULT, FRANCIS (1986) 'Technology Transfer as a means to Build International Networks', *Industrial Marketing and Purchasing*, 1(1) pp. 59–71.

ARNI, VENKATA R. S. (1984) *Evaluation of technology payments*, UNIDO, I.D./W.G. 429/5 (6 September).

ARROW, KENNETH J. (1962) 'Economic Welfare and the Allocation of Resources for Invention', in *The Rate and Direction of Inventive Activity: Economic and Social Factor*, NBER, Princeton University Press, pp. 609–26, reproduced in D. M. Lamberton (ed.) (1971).

ARROW, KENNETH J. (1970) *Essays in the theory of risk-bearing*, North Holland, Elsevier (2nd edn 1974).

ATAMER, TUGRUL (1980) *Choix des partenaires et modalités de transfert international de technologie*, 3rd Cycle Doctorate in Management Science, University of Sociales Sciences of Grenoble, Institut d'Administration des Entreprises.

BAASCHE, JAMES R. and DUERR, MICHAEL G. (1975) *International Transfer of Technology: A worldwide survey of Chief Executives*, The Conference Board, New York.

BARANSON, JACK (1970) 'Technology Transfer through the International Firm', *American Economic Review* (May).

BARANSON, JACK (1978) *Technology and the Multinationals*, Lexington, Mass., Lexington Books.

BARANSON, JACK (1980) *Critique of International Technology Transfer Indicators*, Paper Commissioned by Science Indicators Unit, National Science Foundation (21 January).

BERNARD, GEORGES (1979) 'Transferts inclus dans les prix en économies de marché et en économies planifiées', *Revue d'Economie Politique*, 2, pp. 238–51.

BHAGWHATI, JAGDISH N. (1978) *The New International Economic Order: The North–South Rebate*, Cambridge, Mass., MIT Press.

BIDARD, C. (ed.) (1984) *La production jointe: nouveaux débats*, Paris, Economica.

BIDAULT, FRANCIS (1982) 'L'exportation de technologie par les P.M.I. et le recours aux services', in *Les services: réponse à la crise pour les P.M.I.?*, proceedings of a conference held in Lyon (25 February) Lyon, DATAR/SESAME – ECONOMIE et HUMANISME, pp. 39–52.

BIDAULT, FRANCIS (1986) 'Le Prix des Techniques: des Principes à la Stratégie', thesis for State Doctorate in Management Science, University of Montpellier (12 December).

BIDAULT, FRANCIS and MULLOR, EVA (1984) 'L'exportation de technologie est-elle rentable?', *Le Moniteur du Commerce International*, 608 (21 May).

BIDAULT, FRANCIS and ZERVUDACHI, ANDRÉ (1983) 'Transfert de technologie: fixer le juste prix', *Le Moniteur du Commerce International*, 559 (13 June).

BIZEC, RENÉ-MARIE (1981) *Les transferts de technologie*, Collection Que sais-je, 1915, Paris, PUF.

BLANC, GERARD (1980) *Le contrat international d'équipement industriel – l'exemple algérien*, thesis for state doctorate in Private Law, University of Aix – Marseille III (12 December), Grenoble, Service de reproduction de Thèses.

BONIN, BERNARD (1985) 'Accords contractuels et transferts internationaux de technologie: les études empiriques', The AREPIT Round Table, University of Paris–Dauphine (4–6 September).

BOULDING, KENNETH E. (1971) 'The Economics of Knowledge and The Knowledge of Economics', *American Economic Review*, 56, pp. 1–13, reproduced in D. M. Lamberton (ed.) (1971).

BOWLER, JOHN E. (1980) 'Payments for Technology', *Les Nouvelles*, 4 (XI) pp. 241–7 (December).

BRUNNER, KARL and METZGER, ALLAN H. (eds) (1977) *Optimal Policies Control Theory, and Technology Exports*, Amsterdam, North-Holland, pp. 203–24.

BUCKLEY, PETER J. and CASSON, MARK (1976) *The Future of the Multinational Enterprise*, London, Macmillan (2nd edn 1978).

BUHLER, NICOLAS, CARRIÈRE, DANIEL, PETIT, GÉRARD and SIT-BON, JEAN-CLAUDE (1981). *Les PME françaises et le transfert de technologie dans les P.V.D.*, Cellule de Recherche ESCAE de Marseille – Ministère de l'Industrie.

CARLIER, PAUL (1977) *'L'ingénierie'*, *Problèmes Economiques*, 1537 (7 September).

CAVES, RICHARD E. (1971) 'International Corporations: The Industrial Economics of Foreign Investment', *Economica*, 38, pp. 1–27, reproduced in Dunning (ed.) (1972).

CAVES, RICHARD E. (1974) 'Industrial Organisation', in Dunning (ed.) (1974).

CAVES, RICHARD E., CROOKELL, HAROLD and KILLING, PETER J. (1983) 'The imperfect market for technology licenses', *Oxford Bulletin of Economics and Statistics*, 3, pp. 254–7.

CAVES, RICHARD E. and MURPHY, WILLIAM F. (1976) 'Franchising: firms, markets, and intangible assets', *Southern Economic Journal*, 42(4) (April) pp. 572–86.

CAVES, RICHARD E. and PUGEL, THOMAS A. (1980–2) *Intra industry differences in conduct and performance: viable strategies in U.S. manufacturing industries*, New York University, Graduate School of Business Administration, Monographs Series in Finance and Economics.

CAVES, RICHARD E. and UEKUSA, MASU (1976) *Industrial organization in Japan*, Washington, D.C., The Brookings Institution.

CHEVALIER, JEAN-MARIE (1977) *L'économie industrielle en question*, Paris, Calmann-Levy.

COASE, RONALD H. (1937) 'The Nature of the Firm', *Economica*, 4 pp. 386–405, reprinted in Poshev and Scott (eds) (1980).

CONSEIL et DÉVELOPPEMENT (1980) *Analyse coûts-avantages des opérations de transfert de technologie*, report of the Ministère de l'Industrie (January).

CONTRACTOR, FAROK J. (1981a) *International Technology Licensing: Compensation, Costs, and Negotiation*, Lexington, Mass., Lexington Books.

CONTRACTOR, FAROK J. (1981b) 'The Role of Licensing in International Strategy', *Columbia Journal of World Business* (Winter) pp. 73–9.

CONTRACTOR, FAROK J. and SAGAFI-NEJAD, TAGI (1981) 'International Technology Transfer: Major Issues and Policy Responses', *Journal of International Business Studies* (Fall).

CONTRACTOR, FAROK J. (1984) 'Choosing between direct investment and licensing: theoretical considerations and empirical tests', *Journal of International Business Studies* (Winter) pp. 167–88.

CREMADEZ, DUMONT, HOFFSTETTER (1982) *Maîtrise de l'Effet en Retour des Transferts de Technologie et Stratégie de Développement*, CESA, Jouy-en-Josas, France.

DELACOLLETTE, JEAN (1981) *'Exportation et Transfert de Technologie'*, *Annales des Sciences Economiques Appliquées*, 37 (4) pp. 167–85.

DEMSETZ, HAROLD (1969) 'Information and efficiency: another viewpoint', *Journal of Law and Economics*, 12 (April) (1983) pp. 1–22.

DETRIE, JEAN-PIERRE and RAMANANTSOA, BERNARD (1983) *Stratégie d'Entreprise et Diversification*, Paris, Nathan.

DROUVOT, HUBERT and ECHEVIN, CLAUDE (1985) *Gérer la technologie*, Grenoble, IAE and FNEGE.

DUNNING, JOHN H. (1980) 'Toward an Eclectic Theory of International Production: some empirical tests', *Journal of International Business Studies* (Spring–Summer) pp. 9–31.

DUNNING, JOHN N. (1982) *'Vers une taxonomie des transferts de technologie et de leurs effets en retour pour les pays de l'OCDE – note méthodologique'*, in *Les enjeux des Transferts de Technologie Nord/Sud – études analytiques*, Paris, OECD.

DUNNING, JOHN H. (1983) 'Market Power of the Firm and International Transfer of Technology', *International Journal of Industrial Organization*, 1, pp. 333–51, Amsterdam, North Holland.

DUNNING, JOHN H. (ed.) (1972) *International Investment – Selected Readings*, Harmondsworth, Penguin Books.

DUNNING, JOHN H. (ed.) (1974) *Economic Analysis and Multinational Enterprise*. Winchester, Mass., Allen and Unwin.

DURAN, H. and DROUVOT, H. (1978) *'PME et Transfert de Technologie: les compresseurs Bernard'*, *Enseignement et gestion*, Special issue (May).

DUFOURT, DANIEL (1979) *Transfert de technologie et dynamique des systèmes techniques: éléments pour une politique nouvelle de la recherche technique*, complementary thesis in Economics, University of Lyon II, Saint-Etienne: Conseil et Développement.

EMMANUEL, ARGHIRI (1981) *Technologie appropriée ou technologie sous-développée*, collection perspective multinationale, Paris, PUF–IRM.

EITEMAN, DAVID K. and STONEHILL, ARTHUR I. (1982) *Multinational Business Finance*, Reading, Mass., Addison-Wesley (3rd edn).

FINNEGAN, MARCUS B. and MINTZ, HERBERT H. (1978) 'Determination of a reasonable royalty in negotiating a license agreement: practical pricing for successful technology transfer', *Licensing Law and Business Report*, 1 (2) (June–July) pp. 13–23.

GAUDIN, JACQUES-HENRI (1982) *Stratégie et Négociation des Transferts de Technique*, Paris, Editions du Moniteur.

GERMIDIS, DIMITRI (ed.) (1977) *Le transfert technologique par les firmes multinationales*, Paris, OECD.

GONOT, PIERRE F. (1974) *'Les transferts technologiques'*, AFSE Conference, Lille, quoted by Madeuf (1981).

GRAHAM, E. M. (1982) *'Termes et conditions des transferts de technologie vers les pays en voie de développement'*, in *Les enjeux des transferts de technologie Nord/Sud – études analytiques*, Paris, OECD.

HALL, G. and JOHNSON, J. (1970) 'Transfer of United States Aerospace Technology to Japan', in Vernon (ed.) (1970).

HEE CHEE YONG (1979) *A Negotiation Framework for Technology Importers in Less-Industrialized Countries*, Ph.D. dissertation, University of Washington.

HIRSCHLEIFER, JACK (1956) 'On the Economics of Transfer Pricing', *The Journal of Business*, XXIX (3) (July) pp. 172–84.

HIRSCHLEIFER, JACK (1971) 'The Private and Social Value of Information and the Reward of Inventive Activity', *American Economic Review*, 61, pp. 561–74.

HOUSSIAUX, JACQUES (1957) *'Quasi-Intégration, Croissance des Firmes et Structures industrielles'*, *Revue Economique*, 3, pp. 385–411.

HUARD, PIERRE (1974) *Objectif et système de guidage de l'entreprise*, Editions du CNRS.

HYMER, STEPHEN H. (1976) *International Operations of National Firms: A Study of Direct Foreign Investment*, Boston, Mass., MIT Press.

INSITUTO DO INVESTIMENTO ESTRANGEIRO (1983) *'Pagamentos de technologia e repartiçao de Lucros: um exercicio sobre o metodo UNIDO'*, *Investimento e technologia*, 1.

INTI (1983) 'Research and Development expenditure as a criteria for technology payment evaluation', UNIDO duplicated, ID/WG, 383/7 (11 March).

JACQUEMIN, ALEXIS (1967) *L'entreprise et son pouvoir de marché*, Paris, Presses Universitaires de France.

JACQUEMIN, ALEXIS (1974) *'Pouvoir de l'entreprise et incertitude'*, *Economies et Sociétés*, Cahier de l'ISEA, série M, 28, pp. 145–57.

JANISEWSKI, HUBERT A. and BESSO, MARC (1982) 'Remuneration in Technology Transfer', in *Licensing Law Handbook 1982*, Clark-Boardman.

JOHNSON, HARRY G. (1970) 'The efficiency and welfare implication of the international corporation', in Kindelberger (ed.) (1970).

JOLY, CHRISTIAN (1981) *Bibliographie sur le transfert de technologie*, Paris, Economica (1975) (2nd edn).

KILLING, PETER J. (1975) *Manufacturing under Licence in Canada*, PhD thesis, quoted by Bonin (1985).

KINDELBERGER, CHARLES P. (1969) 'The Theory of Direct Investment', in *American Business Abroad*, New Haven, Yale University Press.

KINDELBERGER, CHARLES P. (ed.) (1970) *The International Corporation: A Symposium*. Cambridge, Mass., MIT Press.

KNICKERBOCKER, FREDERICK T. (1973) *Oligopolistic Reaction and Multinational Enterprise*, Boston, Division of Research, Harvard University.

KOERBER, PHILIPPE (1980) *L'approche allemande du transfert de technologie, l'influence de la coopération publique avec les P.V.D. sur les comportements des P.M.I. face au transfert de technologie: le cas des I.A.A.*, Dissertation for the Diploma of Advanced Studies, University of Lyon II – University of Lyon III – Lyon Graduate School of Business.

KOPITS, GEORGE (1976) 'Intra-firm Royalties Crossing Frontiers and Transfer Pricing Behavior', *The Economic Journal* (December).

LALL, SANJAYA (1976) 'The Patent System and the Transfer of Technology to Less-developed Countries', *Journal of World Trade Law* (January) pp. 1–15.

LAMBERTON, D. M. (ed.) (1971) *Economics of Information and Knowledge*, Harmondsworth, Penguin Books.

LOVELL, ENID BAIRD (1968) *Domestic licensing practices: a survey*, The Conference Board 1968, Experiences in Marketing Management, 18.

LOVELL, ENID BAIRD (1969) *Appraising Foreign Licensing Performance*, Studies in Business Policy, 128, National Industrial Conference Board.

MADEUF, BERNADETTE (1981) *L'ordre technologique international*, Notes et Etudes Documentaires, Paris, La Documentation Française, 4641–4642 (10 November).

MADEUF, BERNADETTE (1985) *'La compétitivité technologique: Evolution au sein de l'OCDE 1970–1980'*, Communication à la Table Ronde de l'AREPIT – University of Paris–Dauphine (5–7 September).

MAGEE, STEPHEN P. (1977) 'Application of the dynamic limit pricing model to the price of technology and international technology transfer', in Brunner and Metzger (eds) (1977).

MAGEE, STEPHEN P. (1978) 'Information and the Multinational Corporation: An Appropriability Theory of Direct Foreign Investment', in Bhagwati (ed.) (1978).

MAHIEUX, FRANCIS (1981) *'Le négoce de licence'*, *Revue Française de Gestion* (January–February) pp. 65–75.

MANSFIELD, EDWIN (1983) 'International Transfer of Technology: an economic analysis', in Remiche (ed.) (1983).

MARCH, J. C. and SIMON, H. A. (1964) *Les Organisations*, Paris, Dunod.

MARCHESNAY, MICHEL (1986) *La Stratégie*, Paris, Chotard Associés.

MARECHAL, C. (1981) *Les paiements dans les contrats de licence de brevets et de communication de savoir faire*, Mémoire de DESS, INPI, Paris.

MAROIS, BERNARD (1980) *Les transferts de technologie internationaux: analyse conceptuelle et étude empirique*, CESA, Cahiers de Recherche 137.

MARTINET, ALAIN CH. (1983) *Stratégie*, Paris, Vuibert.

MASON, R. HAL (1973) 'The Multinational Firm and the Cost of Technology to Developing Countries', *California Management Review* XV (4) (Summer).

MICHALET, CHARLES A. (1976) *Le Capitalisme Mondial*, Paris, Presses Universitaires de France.

MICHALET, CHARLES A. and DELAPIERRE, MICHEL (1976) *The Multinationalization of French Firms*, Chicago, Academy of International Business, quoted by Caves, Crookell and Killing (1983).

MODIANO, GIOVANNA (1983) *'Les contrats de transfert de technologie'*, *Droit et pratique du Commerce International*, Tome 9 (3) pp. 553–91.

MORVAN, YVES (1976) *Economie industrielle*, Paris, Presses Universitaires de France.

MOUGEOT, JEAN-CLAUDE, *'La rémunération des transferts de technologies'*, société TECHNOVA, document multigraphié.

OECD (1981) *Les enjeux des transferts de Technologie Nord/Sud*, Paris.

ORLEANS, GODFREY P. (1981) 'Pricing Licensing of Technology', *Les Nouvelles*, 4 (XVI) (December).

OZANNE, U. B. and HUNT, S. D. (1971) *'The Economic Effects of Franchising'*, US Senate, Select Committee on Small Business, Committee Print, 92 cong., 1 sess., Washington: Government Printing Office, quoted by Caves and Murphy (1976).

OZAWA, TERUTOMO (1981) 'Technology Transfer and Control Systems: The Japanese experience', in Sagafi-Nejad, Moxon and Perlmutter (1981).

PENROSE, EDITH (1973) 'International Patenting and the Less-developed Countries', *The Economic Journal* (September) pp. 768–85.

PERLMUTTER, HOWARD V. and HEENAN, DAVID A. (1986) 'Cooperate to compete globally', *Harvard Business Review* (March–April) pp. 136–41.

PERLMUTTER, HOWARD V. and SAGAFI-NEJAD, TAGI (1981) *International Technology Transfer: guidelines, codes and a muffled quadrilogue*, New York, Pergamon Press.

PIATTI, MARIE-CHRISTINE (1980) *'La protection et la valorisation du savoir faire'*, in Drouvot and Echevin (1985).

PERLITZ, MANFRED (1980) 'Compensatory Arrangements in International Licensing Agreements: effects on markets penetration abroad, foreign pricing policy, and conflict areas between Licensee and Licensor', *Management International Review*, 20 (1) pp. 75–82.

PERRIN, JACQUES (1983) *Les transferts de technologie*, Paris, La Découverte, Maspero.

PERROUX, FRANÇOIS (1973) *Pouvoir et Economie*, Collection Etudes, Paris, BORDAS.

PORTER, MICHAEL E. (1980) *Competitive Strategy: Techniques for Analyzing Industries and Competitors*, New York, The Free Press.

PORTER, MICHAEL E. (1985) *Competitive Advantage: Creating and Sustaining Superior Performance*, New York; The Free Press.

POSNER, RICHARD A. and SCOTT, KENNETH E. (eds) (1980) *Economics of Corporation Law and Securities Regulation*, Boston, Little, Brown & Co.

RADNER, ROY (1979) 'The Economic Value of Information', paper presented at the 60th Anniversary Meeting of the International Union of Radio Science Brussels (17–18 September) p. 3–1.

RAGAZZI, GEORGIO (1973) 'Theories of the Determinants of Direct Foreign Investment', *IMF Staff Papers* (July) pp. 471–98.

REMICHE, BERNARD (ed.) (1983) *Transfert de technologie: Enjeux économiques et Structures juridiques*, Paris – Louvain, CABAY-ECONOMICA.

RIVELINE, CLAUDE (1983) *'Nouvelles approches des processus de décision: les apports de la recherche en gestion'*, *Revue Futuribles*. 72 (December).

ROOT, FRANKLIN R. (1981) 'The pricing of international technology transfers via non-affiliate licensing agreements', in Sagafi-Nejad, Moxon and Perlmutter (1981).

ROOT, FRANKLIN E. and CONTRACTOR, FAROK J. (1981) 'Negotiating Compensation in International Licensing Agreements', *Sloan Management Review*, 22 (2) (Winter) pp. 23–32.

ROTHSCHILD, WILLIAM E. (1984) *How to gain (and maintain) the competitive Advantage in Business*, New York, McGraw-Hill.

RUGMAN, ALAN M. (1980) 'Internalization as a General Theory of Foreign Direct Investment: A Re-Appraisal of the Literature', *Weltwirtschaftliches Archiv*, CXIV (June) pp. 365–79.

RUGMAN, ALAN M., LECRAW, DONALD J. and BOOTH, LAURENCE D. (1985) *International Business: Firm and Environment*, New York McGraw-Hill series in Management.

SAGAFI-NEJAD, TAGI (1979) *Developmental Impact of Technology Transfer: theory, determinants and verifications from Iran, 1954–74*, Ph.D. dissertation, University of Pennsylvania.

SAGAFI-NEJAD, TAGI and MAGEE, STEPHEN P. (1981) *Foreign Direct Investment Theory and Capital Theory considerations in the Pricing of Technology: an Empirical Investigation*, working paper 81/82–5–6, University of Texas at Austin, Graduate School of Business, Department of Marketing Administration.

SAGAFI-NEJAD, TAGI, MOXON, RICHARD and PERLMUTTER, V. (1981) *Controlling International Technology Transfer: Issues, Perspectives and Policy implications*, New York, Oxford, Pergamon Press.

SAVARY, JULIEN (1981) *Les Multinationales Françaises*, Collection perspective multinationale, Paris, PUF–IRM.

SCHERER, FREDERICK M. (1980) *Industrial Market Structure and Economic Performance*, Chicago, Rand McNally (2nd edn).

SILBERSTON, A. (1967) 'The patent system', Lloyds Bank Review, 84, pp. 32–44, reprinted in Lamberton (ed.) (1971).

SOEWONDO, HARTONO (1983) *Acquisition de technologie étrangère par les petites et moyennes industries en Indonésie*, dissertation for the Diploma of Advanced Studies, University of Lyon III, Lyon Graduate School of Business (3 October).

SRAFFA, PIERO (1926) 'The Laws of Returns under Competitive Conditions', *The Economic Journal*, XXXVI, pp. 535–50.

STEEDMAN, IAN (1984) *'L'importance empirique de la production jointe'*, in Bidard (ed.) (1984).

TEECE, DAVID J. (1981) 'The Market for Know-How and the Efficient International Transfer of Technology', *Annals of the American Academy of Political and Social Science*, 458 (November).

TEECE, DAVID J. (1976) *The Multinational Corporation and the Resource Cost of International Technology Transfer*, Cambridge, Mass., Ballinger Publisher Co.

TELESIO, PIERO (1979) *Technology Licensing and Multinational Enterprise*, New York, Praeger.

UNIDO (1982) *Principes directeurs pour l'évaluation des accords de transfert de technologie*, série 'Mise au point et transfert des techniques', 12, New York, United Nations.

UNIDO (1983a) *Technological Information Exchange System (TIES), Technology Payments and Profit Sharing in Portugal*, I.D./W.G. 383/6 (14 February).

UNIDO (1983b) *Technology payments evaluation: Summary results of a Pilot exercise*, 8th meeting of Heads of Technology Transfer registries, Caracas, Venezuela (17–20 October) ID/WG (31 August) (1981).

VAITSOS, CONSTANTINE V. (1974) *Intercountry Income Distribution and Transnational Enterprise*, Clarendon Press–Oxford University Press (2nd edn 1976).

VAUCHER, MAURICE (1978) *'Comment exporter la matière grise'*, Entreprises Rhône-Alpes (November).

VENNIN, BRUNO and CONSEIL ET DÉVELOPPEMENT (1979) *Petites et moyennes Entreprises Industrielles et Transfert de Technologie – Pratique du Transfert et Internationalisation*, Rapport au CORDES.

VERNON, RAMOND (1960) 'International Investment and International Trade in the Product cycle', *Quarterly Journal of Economics*, 80 (20) (May) pp. 190–207.

VERNON, RAYMOND (1979) 'The product cycle hypothesis in a new international environment', *Oxford Bulletin of Economics and Statistics* (November).

VERNON, RAYMOND (ed.) (1970) *The Technology Factor in International Trade*, New York, Columbia.

WEIL, ALAIN (*Rapport d'un groupe de travail animé par*) (1980) *Les transferts de technologie aux pays en voie de développement par les petites et moyennes industries*, Paris, La Documentation Française.

WILLIAMSON, OLIVER (1975) *Markets and Hierarchies: analysis and antitrust implications*, New York, The Free Press.

YONG, SE-JUNG and LASSERRE, PHILIPPE (1982) '*La modalité de l'acquisition de technologie dans les entreprises de pays en voie de développement*', *Revue Français de Gestion* (September–October).

ZERVUDACHI, ANDRÉ (1983) *Le prix et la rémunération des techniques exportées*, Mémoire pour l'Atelier de Recherche en Environnement, ESC Lyon (April).

Index